COINTEGRATION

Also by B. Bhaskara Rao

THE ECONOMETRICS OF DISEQUILIBRIUM MODELS

Cointegration

for the Applied Economist

Edited by

B. Bhaskara Rao
Associate Professor in Economics
University of New South Wales
Kensington, Australia

St. Martin's Press New York

COINTEGRATION

For information, address:

St. Martin's Press, Scholarly and Reference Division,
175 Fifth Avenue, New York, N. Y. 10010

First published in the United States of America in 1994

Printed in Great Britain

ISBN 0-312-12177-6 (cloth)
ISBN 0-312-15809-2 (paperback)

Library of Congress Cataloging-in-Publication Data
Cointegration : for the applied economist / edited
by B. Bhaskara Rao.
p. cm.
Includes bibliographical references and index.
ISBN 0-312-12177-6 (cloth)
ISBN 0-312-15809-2 (paperback)
1. Economics—Statistical methods. 2. Correlation (Statistics)
I. Bhaskara Rao, B., 1939-
HB137.C65 1994
330—dc20 94-5981
CIP

CONTENTS

List of Tables

DATA APPENDIX TABLES

List of Figures

PREFACE

Economists and especially macro economists are confronted now with a new challenge by the econometricians. Time series econometricians have pointed out that many macroeconomic variables are non–stationary and therefore conventional estimation techniques, based on the usual classical assumptions about the properties of the disturbance terms, lead to mistaken inferences. Further developments in time series analysis have shown that if the variables in a model are non–stationary, cointegration and error correction techniques should be used to estimate such models.

Subsequently there has been an outburst of highly technical papers in the learned journals on various methods of testing for unit roots and estimating cointegrating relationships and error correction specifications. However the usual space constraints imposed by the editors of these journals seem to have forced various authors to be rigourous in exposition and abstract from the pedagogic aspects of their contributions. Consequently much of the literature on unit roots and cointegration has remained inaccessible to the majority of the applied economists.

This volume of expository essays aims to fill this gap and reduce the usual long time lag between the development of new techniques and their application in the day to day applied economic research. We have selected for inclusion into this volume important papers which have been originally written with a pedagogic slant by some well known specialists. However most of these papers have been now updated by their authors to make them even more pedagogic. All the authors have also supplied the data used in their contributions and these are given in the data appendix. Readers can use these data to replicate the original results to gain experience and confidence before various techniques are used in their own research.

Originally it was our intention to supply all the data on a floppy disk with this book. However that would have significantly added to its price. Therefore we have decided to make these data available through the usual international computer networks and further details are given in the data appendix.

These expository essays will also be highly useful to the undergraduate and graduate students in the economics, business and finance degrees. Virtually every essay can be assigned as a required reading in the applied econometrics courses. Replications of various results, using the data in the appendix, will strengthen students' practical econometric and quantitative skills. Therefore this book will be a valuable source of project topics to the teachers of these courses.

We must record our deep sense of gratitude to all the contributors to this volume. Their response to our initial request to revise their papers to make them more pedagogic was immediate and spontaneous. We thank the editors and publishers of the *Federal Reserve Bank of St. Louis Review* and the *American Economic Review* for permission to use some published material. We are grateful to Dr. Giovanna Davitti, Commissioning Editor of the Macmillan Press, for her encouragement, patience and guidance. Thanks are also due to Mrs. Naroj Nafratitti and Mrs. Mayda Shahinian of the School of Mathematics, University of New South Wales, for skilfully typing the manuscript; to Ms. Belinda Allen and Ms. Teresa Attfield of the Audio Visual Unit, University of New South Wales, for drawing all the figures in this book and to Mr. Tarlok Singh of the Reserve Bank of India and Mr. Roger Perman of the University of Strathclyde for help with proof reading. My final thanks go to Professor V. K. Srivastava of the University of Lucknow for his constant encouragement and inspiration.

B.B.R.

NOTES ON THE CONTRIBUTORS

David Dickey is Professor of Statistics at the North Carolina State University. He received his Ph.D. from Iowa State University and is well known for his contributions on testing for unit roots in time series. His work has been published in well known journals such as the *Journal of the American Statistical Association, American Statistician, Econometrica, Biometrika, Journal of Business and Economic Statistics, Annals of Statistics* and *the Review of the St. Louis Federal Reserve Bank.*

Darryl Holden is Lecturer in Economics at the University of Strathclyde, Glasgow. He graduated from the Universities of Manchester and London. His research interests are in the areas of econometrics and regional economics.

Dennis Jansen is Associate Professor in Economics at the Texas and A & M University and a research fellow at the Private Enterprise Research Centre. He obtained his Ph.D. from University of North Carolina. His research interests are in the fields of applied macroeconometrics and monetary theory and policy.

Yash Mehra is a vice president and economist at the Federal Reserve Bank of Richmond. He graduated from the Punjab University and the Delhi School of Economics and obtained his Ph.D. from the University of Minnesota. He published papers on monetary economics in the *American Economic Review, Journal of Political Economy, Review of Economics and Statistics, Journal of Money, Credit and Banking, Southern Economic Journal, Journal of Macroeconomics* and *Economic Review of Federal Reserve Bank of Richmond.*

Glenn Otto is Lecturer in Economics at the University of New South Wales, Sydney. He received his Ph.D from the Queens' University. His research interests are in the areas of macroeconomics and applied econometrics. He published papers in *Economics Letters, Journal of International Money and Finance* and *Australian Economic Papers.*

Roger Perman is Lecturer in Economics at the University of Strathclyde, Glasgow. He was educated at the universities of Manchester, Leicester and London. His research interests and publications are in the fields of applied econometrics and environmental economics.

Pierre Perron is Professor of Economics at the Université de Montréal and a research associate at the Centre de recherche et développement en économique. He previously held faculty positions at Princeton University. His Ph.D. in economics is from the Yale University and he has published papers in well known journals such as the *Econometrica, Econometric Theory, Journal of Econometrics, Biometrika, Journal of Business and Economic Statistics* and *Journal of Economic Dynamics and Control.*

Bhaskara Rao is Associate Professor in Economics at the University of New South Wales, Sydney. He graduated from the Banaras Hindu University and the London School of Economics and received his Ph.D. from the University of New South Wales. He published papers on macroeconomics in journals such as the *Economic Journal, Review of Economics and Statistics, Economics Letters, Southern Economic Journal, Australian Economic Papers, Journal of Macroeconomics, Applied Economics* and *Journal of Quantitative Economics.* He is the co-author of the *Econometrics of Disequilibrium Models* (1990).

Daniel Thornton is an assistant vice president and research economist at the Federal Reserve Bank of St. Louis. He joined the research department of the Federal Reserve Bank of St. Louis in 1981 as a staff economist and was promoted to his current position in 1989. He received his Ph.D. from the University of Missouri-Columbia in 1976. He published papers in the *Federal Reserve Bank of St. Louis Review* and in other professional journals. He is an Associate Editor of the *Journal of International Financial Markets, Institutions and Money.*

1 Editor's Introduction

B. Bhaskara Rao

1.1 INTRODUCTION

Methodological revolutions in economics are not new. Economics has gone through the Keynesian, monetarist, neoclassical and new classical revolutions in the post war period. The major impact of each one of these revolutions is that they call for a fundamental change in our way of thinking about modelling economic phenomena. Such revolutions in economics are invariably controversial partly because they often imply that existing policy measures are inappropriate and should be abandoned in favour of a new set of policies. Herein lies a major source of controversy. The old and new policies have different adherents depending on their sense of economic fairness and justice. Therefore it is hard to derive widely acceptable conclusions about the relative merits of these revolutions using a positive framework.

At the moment, both economics and econometrics are going through a new type of revolution: the unit roots and cointegration revolution. This new revolution calls for a fundamental change in thinking about methods of estimation of economic relationships as well as modelling fluctuations in economic activity. However, there is a major difference between the new and the old revolutions. Debates about the significance of the new revolution can be conducted within a positive framework and without the need for its appeal to one's sense of economic justice. Consequently the new revolution has attracted not only the attention of specialist econometricians but also a large number of policy oriented applied economists. Methodological predilections are less important for judging the significance of the new revolution!

In what follows, we shall first provide an intuitive and brief, explanation of the significance of unit roots and cointegration; give a few guidelines to some background material to understand the rapidly accelerating literature on unit roots and cointegration and finally provide an overview of the essays in this volume.

1.2 UNIT ROOTS AND COINTEGRATION

The standard classical methods of estimation, which we routinely use in the applied econometric work, are based on the assumption that the means and variances of the variables are well defined constants and independent of time. However, applications of the unit root tests have shown that these assumptions are not satisfied by a large number of macroeconomic time-series. Variables whose means and variances change over time are known as non-stationary or unit root variables. Furthermore the unit root revolution has also shown that using classical estimation methods, such as the ordinary least squares (OLS), to estimate relationships with unit root variables gives misleading inferences. This is known as the spurious regression problem and an intuitive explanation of its significance is as follows. If the means and variances of the unit root variables change over time, all the computed statistics in a regression model, which use these means and variances, are also time dependent and fail to converge to their true values as the sample size increases. Furthermore conventional tests of hypothesis will be seriously biased towards rejecting the null hypothesis of no relationship between the dependent and independent variables. This is a serious problem if the null hypothesis is true.[1]

The economic implications of the unit roots literature are also equally profound. So far alternative macroeconomic paradigms have treated economic fluctuations as temporary deviations from a stable trend rate of growth of output and offered different explanations for these fluctuations. Therefore while there is bound to be some disagreement between these theories on the merits of alternative short-run stabilisation policies, such disagreements are less noticeable on policies to promote the trend growth rate of output.

1. The spurious regression problem has other implications. Phillips (1986) developed a formal model for regressions between unit root variables to show, for example, that the $D.W$ statistic converges towards zero. This is an important finding because low $D.W$ statistics often indicate that the variables in a regression model are non-stationary.

Unit root tests, on the other hand, have shown that the assumption about the stability of the long-run trend rate of growth of output is untenable because aggregate output in many countries is found to be non-stationary. This finding casts a doubt on the usefulness of the existing business cycle theories in which a distinction is made between the determinants of the trend rate of growth of output and cycles.

So far we have considered some implications of unit roots. However cointegration has also important implications. In this context it is convenient to view cointegration as a technique to estimate the equilibrium or long-run parameters in a relationship with unit root variables. Together unit roots and cointegration, however, have important implications for the specification and estimations of dynamic economic models.

While much of the existing neo and new classical theories use the equilibrium framework, it is generally believed that the Keynesian theory is based on the disequilibrium framework. This does not, however, imply that cointegration is useful only for the estimation of the neo and new classical models. If a distinction is made between equilibrium as merely a state of rest and a state of rest which is also an optimal state, then Keynesian economics can be interpreted as equilibrium economics, albeit an equilibrium in which resources are under employed. Consequently the usefulness of cointegration techniques to the Keynesian models should not be underestimated.

What is new in cointegration is that it calls for a different strategy to model dynamic models. In the exiting strategies the methodological conflict between the equilibrium framework of the theory and the disequilibrium environment from which data are gathered is resolved by extending the equilibrium specifications to include disequilibrium adjustment mechanisms. The extended equation is then estimated from which estimates of the long-run or equilibrium parameters are derived by imposing equilibrium conditions. These equilibrium parameters are then used to test the underlying theory.

Even though the traditional approach can be criticised for neglecting the problems caused by the presence of unit root variables, the main advantage of cointegration is that it can be used directly to test or falsify the underlying theory. Furthermore cointegration enables utilisation of the estimated long-run parameters into the estimation of the short-run disequilibrium relationships. There is, however, a trade-off. Estimation methods of cointegrating regressions and joint estimation of both the long and short-run specifications are computationally demanding. But this difficulty should not mask the significance of the unit roots and cointegartion revolution.

1.3 SOME GUIDELINES

For several years economists and econometricians have paid relatively less attention to the statistical theory underlying the time-series models. Therefore the theory underlying unit roots and cointegration was not quickly understood and appreciated. Furthermore there is a paucity of expository material on these developments for the applied economists. One of the objectives of this volume is to fill this gap and also reduce the long lags between published research and its application in applied econometric work.

At this juncture a few words about some background readings and computer software packages would be appropriate. Without some rudimentary background in time-series models and acquaintance with the colourful terminology used to describe various time-series models it is hard to follow even a simple discussion on unit roots and cointegration.

The reader is advised to first revise his or her knowledge of elementary econometric methods and time-series models. For example Maddala (1992) has an excellent treatment of various topics on econometric methods and most applied economists will find this book easy to understand. A comprehensive exposition of time-series models is given by Pindyke and Rubinfeld (1991). Both books also have excellent introductory chapters on unit roots, cointegration and error correction models; see also Cuthbertson (1989, pp. 256-271) for an interesting discussion of econometric modelling with time-series data.

Several well known statistical packages can be used to test for unit roots, estimate cointegrating regressions and error correction specifications. However, the reader is advised to make sure from the manuals of his or her favourite package that at least some important results pertaining to unit roots and cointegration can be obtained in a straight forward manner, perhaps by simply pressing an additional button on the computer keyboard. For example *Microfit, PcGive, RATS, SAS, TSP* etc. are all widely used in the applied econometric work. It is not an exaggeration to say, using the parlance of the computer buffs, that *Microfit,* and *RATS* have attained the status of maniac programmes! However, all these packages have excellent subroutines to obtain various results on unit root and cointegration tests. Among the authors of this volume Dickey, Jansen and Thornton have used *SAS,* both Mehra and Perron *RATS* and Holden and Perman and Otto *Microfit. Microfit* has also some excellent menus for post-regression diagnostic test statistics.

The order in which the essays in this book can be read and understood depends partly on one's econometric background. Readers familiar with the literature on the time-series models and perhaps have used some unit root tests in their applied work can read straight away the essays by Dickey, Jansen and Thornton, Mehra and Perron. On the other hand readers with a modest background in econometric methods are advised to read first the references cited earlier and then Otto, Holden and Perman and Mehra.[2] Dickey, Jansen and Thornton can then be read easily. Finally prior to reading the essay by Perron, it is advisable to read Duck (1992) for an overview of the significance of Perron's contribution.

1.4 AN OVERVIEW

There are broadly four major steps in applying unit root and cointegration techniques. First, unit root tests are applied to determine if the variables in a regression are stationary or nonstationary. Second, cointegrating regressions are estimated if the variables satisfy certain conditions. These cointegration regressions are the long-run or equilibrium relationships between these variables. It should be, however, noted that these relationships may not have a causal interpretation. Therefore theoretical insights are necessary to determine what causes what. Third, the short-run or the dynamic disequilibrium relationships are estimated utilising the estimates of the long-run parameters within

2. If some readers need a more elementary exposition than Holden and Perman they should consult Perman (1991).

the error correction framework. Finally, the robustness of the estimated dynamic disequilibrium relationships is determined by subjecting them to the standard diagnostic tests.

Sometimes it might be possible to show that the unit root null hypothesis can be rejected for a set of variables. Needless to say this would considerably simplify estimation because all the standard classical methods can be utilised in these circumstances. At the moment there seem to be three promising alternative approaches on these lines. Firstly, the null and alternative hypotheses in the standard unit root tests can be interchanged. It is well known that the power of these tests, against the alternative to the null, is low. Secondly, a Bayesian approach can be used in these tests because its power is symmetric with respect to the null and alternative hypotheses. Thirdly, the unit root null hypothesis can be confronted with a stronger alternative hypothesis. In our view the third alternative, proposed by Perron (1989), seems to be most promising; see Rao (1993a, b) for applications of the Perron (1989) tests.

The aforesaid overview is a highly simplified and condensed version of the major steps involved in applying unit root and cointegration techniques. In practice, however, the applied economist will encounter several problems. Firstly, there is more than one method of testing for the presence of unit roots and to estimate the parameters in the cointegrating regressions. Secondly, the critical values for each of these tests and estimation methods depend upon the number of variables and also the specification selected for an equation. Nonetheless it can be said that the augmented Dickey-Fuller test (ADF) is widely used to test for the presence of unit roots in the variables. A more powerful alternative is the Phillips and Perron non-parametric test. The Engle and Granger two-step method is the simplest method of estimation of the cointegrating regression and error-correction formulation. However, the Johansen maximum likelihood method is now widely used for this purpose.

The papers in this volume adequately explain these standard and frequently used procedures. Furthermore various other alternative methods are considered and their relative merits are evaluated by applying them to real the world data. Each paper starts with a comprehensive introduction. For the convenience of the reader we shall provide a brief outline of the salient features of these papers.

In their paper Dickey, Jansen and Thornton use the Dickey-Fuller and the augmented Dickey-Fuller tests to test for unit roots in the variables of the demand for money function. Both tests seem to yield similar conclusions. Next they explain how the cointegrating regressions can be estimated using three different approaches viz. the Engle and Granger two-step, the Johansen ML, and the Stock-Watson procedures. The Johansen ML method seems to yield more satisfactory results. A detailed step-by-step explanation of the Johansen ML and the Stock-Watson methods is given in the Appendix to Chapter 2.

Holden and Perman's paper is somewhat similar in scope to the Dickey, Jansen and Thornton paper illustrating the usefulness of the Phillips and Perron non-parametric test. The discussion of various unit root tests is very comprehensive and an easy to follow step-by-step sequential procedure to conduct these tests is given. The error-correction formulation is discussed in some detail. Although the Johansen ML method is used, there is a detailed discussion of the application of cointegration in econometric modelling. Some of these tests are applied to the variables of the consumption function in the Appendix to Chapter 3.

We have already alluded to the usefulness of Perron (1989). He had shown that if the unit root null is tested against the alternative that the trend function has undergone a one-time structural change, the unit root null can be rejected for eleven out of the fourteen variables in Nelson and Plosser's (1982) seminal study. However, Perron (1989) assumed that the date at which the trend function had undergone a structural change is known *a priori.* In his current paper Perron extends his earlier work so that the date of the structural change can be determined by data. These new tests are applied to a set of historical international data on *GDP* of eleven countries. He shows that there is no major conflict between the conclusions based on his earlier and new methods. However, his new tests reject the unit root hypothesis for seven out of eleven countries. In the four countries for which the null was not rejected the test statistics did not yield consistent estimates of the break point.

Mehra's paper is a good example of the application of the unit root and cointegration techniques in the day-to-day applied econometric work. In addition Mehra also illustrates the usefulness of these techniques in the Granger-Causality tests. He analyses the relationships between wage-price inflation and finds that while wages are adjusted for inflation, the usual presumption, underlying the markup equation, that price are set as a markup on unit costs is not tenable for the U.S. economy. His results are very robust and his paper should inspire many applied economists.

Finally Otto's paper is on a different but closely related topic. Otto abstracts from the unit root and cointegration aspects of his paper to focus on various diagnostic tests. These tests are important for determining the adequacy of the final results obtained from the application of unit root and cointegration techniques. The final dynamic disequilibrium equation, based on the error-correction formulation, should be subjected to a battery of diagnostic tests before it is considered to be adequate. Otto illustrates his tests using the demand for money for Canada as an example. It would be interesting to use Otto's data and some of the techniques discussed in the other papers to obtain a satisfactory error-correction formulation of the Canadian demand for money. The reader may try this as an exercise.

1.5 CONCLUDING OBSERVATIONS

The table of contents is exhaustive and should help to quickly locate where a topic of interest is discussed in the book. The index is thorough but by no means exhaustive. It is impossible to list every citation of frequently cited names like Dickey and Fuller, Phillips and Perron, Engle and Granger and concepts such as the augmented Dickey-Fuller test, Phillips-Perron tests, the Johansen maximum likelihood method etc. Nonetheless an attempt has been made to include all the important concepts, contributions and names in the index. Chapter numbers are prefixed to the section and sub-section numbers. 3.2.1 means that it is subsection 1 of section 2 of chapter 3. Similarly all the tables and figures start with the relevant chapter number. For example Table 4.2 is Table 2 in chapter 4. However, chapter numbers are not prefixed to equations because none of the chapters refer to the equations in another chapter.

2 A Primer on Cointegration with an Application to Money and Income*

David A. Dickey, Dennis W. Jansen and Daniel L. Thornton

2.1 INTRODUCTION

For some time now, macroeconomists have been aware that many macroeconomic time-series are not stationary in their levels and that many time-series are most adequately represented by first differences.[1] In the parlance of time-series analysis, such variables are said to be integrated of order one and are denoted $I(1)$. The level of such variables can become arbitrarily large or small so there is no tendency for them to revert to their mean level. Indeed, neither the mean nor the variance is a meaningful concept for such variables.

Nonstationarity gives rise to several econometric problems.[2] One of the most troublesome stems from a common prediction of macroeconomic theory that there should be a stable long-run relationship among the levels of certain economic variables. That

---

*. This paper is based on Dickey, Jansen and Thornton (1991).

1. That is, formal statistical tests often cannot reject the null hypothesis of a unit root. The results of these tests, however, are sensitive to how the tests are performed that is, whether an MA or AR data generating processes is assumed, as in Schwert (1987), and whether the test is performed using classical or Bayesian statistical inference; see Sims (1988) and Sims and Uhlig (1988). These sensitivities are partly due to the lack of power these tests have against an alternative hypothesis of a stationary but large root.

2. It can give rise to the possibility of a spurious relationship among the levels of the economic variables. Also, the parameter estimates from a regression of one such variable on others are inconsistent unless the variables are cointegrated.

is, theory often suggests that some set of variables cannot wander too far away from each other. If individual time-series are integrated of order one, however, they may be "cointegrated". Cointegrations of these variables is stationary even though individually they are not. If these variables are cointegrated, they cannot move "too far" away from each other. In contrast, a lack of cointegration suggests that such variables have no long-run link; in principle, they can wander arbitrarily faraway from each other.[3]

This article illustrates the salient features of cointegration and tests for cointegration. The discussion, initially motivated by the simple example of Irving Fisher's "equation of exchange", draws an analogy between cointegration and unit roots on the one hand and tests for cointegration among multiple time-series and the usual tests for unit roots in univariate time-series analysis on the other. The article then addresses the broader question of the economic interpretation of cointegration by contrasting it with the usual linear, dynamic, simultaneous equation model which is frequently used in macroeconomics.

The article goes on to compare three recently proposed tests for cointegration and outlines the procedures for applying these tests. An application of these tests to U.S. time-series data using alternative monetary aggregates, income and interest rates suggests that there is a stable long-run relationship among real output, interest rates and several monetary aggregates, including the monetary base.

2.2 TESTING FOR COINTEGRATION: A GENERAL FRAMEWORK

Because of the close correspondence between tests for cointegration and standard tests for unit roots, it is useful to begin the discussion by considering the univariate time-series model

$$y_t - \mu = \varrho(y_{t-1} - \mu) + e_t, \tag{1}$$

3. At the present time, tests for cointegration deal only with looking for stable linear relationships among economic variables. Consequently, a failure to find cointegration does not necessarily mean that there is no stable long-run relationship among the variables. It only suggests that there is no stable long-run linear relationship among them.

where y_t denotes some univariate time-series, μ is the mean of the series and e_t is a random error with an expected value of zero and a constant, finite variance. The coefficient ϱ measures the degree of persistence of deviations of y_t from μ. When $\varrho = 1$, these deviations are permanent. In this case, y_t is said to follow a random walk – it can wander arbitrarily far from any given constant if enough time passes.[4]

In fact, when $\varrho = 1$ the variance of y_t approaches infinity as t increases and μ the mean of y_t is not defined. Alternatively, when $|\varrho| < 1$, the series is said to be mean reverting and the variance of y_t is finite.

Although there is a similarity between the tests for cointegration and unit roots, as we shall see below, these tests are not identical. Tests for unit roots are performed on univariate time-series. In contrast, cointegration deals with the relationship among a group of variables, where (unconditionally) each has a unit root.

To be specific, consider Irving Fisher's important equation of exchange, $MV = Pq$, where M is a measure of nominal money, V is the velocity of money, P is the overall level of prices and q is real output.[5] This equation can be rewritten in natural logarithms as:

$$\ln M + \ln V - \ln P - \ln q = O. \tag{2}$$

In this form, the equation of exchange is an identity. The theory of the demand for money, however, converts this identity into an equation by making velocity a function of a number of economic variables; both the form of the function and its arguments change from one theoretical specification to another. In the theory of money demand, V is unobservable and in applied work it is proxied with some function of economic variables,

4. That is, for any numbers $C > 0$ and $0 < p < 1$ and for any starting value Y there is a time, T, such that, for all $t > T$, $\Pr(|Y_t| > C) > p$. When $|\varrho| < 1$ the process generating Y_t is stationary in that it does not wander too far from its mean, i.e. for any given probability p we can find a constant $C > 0$ such that $\Pr(|Y_t - \mu| < C) < p$.

5. The cointegrating vector could be different from the hypothesized one for other reasons as well. For example, money holders might have a money illusion or money demand might not be homogeneous of degree one in real income.

$V^\star$, $\ln V^\star = \ln V + E$, where E denotes a random error associated with the use of the proxy for V.[6] The proxy is a function of one or more observed variables, other than income and prices, that are hypothesized to determine the demand for money. Hence, equation (2) is replaced with

$$\ln M + \ln V^\star - \ln P - \ln q = E. \qquad (2')$$

If the proxy is good, the expected value of E should be zero. Furthermore, E should be stationary, so that, $V^\star$ might deviate from its true value in the short-run. Failure to find a stationary relationship among these variables (that is, to find that they are not cointegrated) implies either that $V^\star$ is a poor proxy for V or that the long-run demand for money does not exist in any meaningful sense.

In essence, the Fisher relationship embodies a long-run relationship among money, prices, output and velocity. In particular, it hypothesizes that the cointegrating vector $(1, 1, -1, -1)$ exists. This vector combines the four series into a univariate series, E. Given this assumption, a test for cointegration can be performed by applying any conventional unit root test to E.

Using conventional unit root tests to test for cointegration, such as the tests preposed by Dickey and Fuller (1979) and Phillips (1987), however, requires prior knowledge of the cointegrating vector. And most often, this vector is unknown. Therefore, some linear combination of these variables, for example,

$$b_1 \ln M + b_2 \ln V^\star + b_3 \ln P + b_4 \ln q \qquad (3)$$

is hypothesized to be stationary, where the cointegrating vector (b_1, b_2, b_3, b_4) is unknown and must be estimated.

2.2.1 Locating Stationary Linear Combination of Variables

Engle and Granger (1987), Stock and Watson (1988), and Johnasen (1988) have suggested alternative tests for cointegration and methods for estimating the cointegrating vectors. While differing in a number of respects, all of these procedures involve locating the "most stationary" linear combinations (among all

6. For the classic discussion of velocity and a long list of its potential determinants, see Friedman (1956). Empirical proxies for velocity often contain one or more of these determinants.

of the possible ones) of the vector time-series in question. If the linear combinations being compared are not chosen *a priori*, but are determined by choosing among all possible vectors, tests for cointegration encounter the type of distributional problems associated with order statistics and multiple comparisons. Hence, it is useful to discuss some of these problems in more detail.

In multiple comparison tests, an experimenter is usually concerned with, say, comparing the highest and lowest sample means among several. He wants to find the pair of sample means with the largest difference to see whether the difference is statistically significant. When the means to be compared are chosen ahead of time, tests for a significant difference between the means can be done using the usual t-statistics. If, however, the means to be compared are chosen simply because they are the largest and smallest from a sample of, say, five means, the rejection rate under the null hypothesis will be much higher than that implied by the percentile of the t-distribution. In order to control for the experimentwise error rate, as it is called, tables of distributions of highest mean minus lowest mean (standardized) have been computed for the case of no true differences in the population means. These "tables of the studentized range" are then used to test for significant differences between the highest and lowest means.[7]

The price paid for controlling the experimentwise error rate is a loss in power. That is, the difference between the means must be much larger than in the case of the standard t-test before it can be declared significant at some predetermined significance level. Thus, the power of the test to detect significant differences is reduced.

In an analogous way, it is difficult to reject the null hypothesis that there are no stationary linear combinations when the observed data are used to estimate the most stationary-looking linear combination before testing for cointegration. This loss of power is evident in the test tables given by Stock and Watson or Johansen, where the percentiles are shifted far away from those of the standard unit root distribution of Dickey and Fuller (1979).[8] Consequently, detecting cointegrating relationships among vari-

---

7. See Steel and Torrie (1980), p. 588, for these tables.

8. See Stock and Watson (1988) and Johansen (1988) for further discussion of the relationship to order statistics.

ables is relatively hard. Some power can be gained, however, if economic theory is used to assign values to some coefficients, *a priori*. Indeed, if theory fully specified the cointegrating vector, as in our example of the Fisher equation, using conventional unit root tests to test for cointegration would be appropriate.

2.2.2 Multiple Cointegrating Vectors

Until now, the possibility that only one linear combination of variables is stationary has been considered; however, this need not be the case. In cases where more than two time-series are being considered, more than one stable linear combination can exist. Also, until now, cointegration was discussed without any explicit reference to a dynamic specification of the levels of the economic variables. Nevertheless, the fact that cointegration is related to their dynamic specification was implicit in the fact that all of the univariate series are $I(1)$. While a number of alternative multivariate representations could be used, it is convenient to use the following multivariate $AR(1)$ representation:

$$Y_t = AY_{t-1} + \epsilon_t, \tag{4}$$

where Y_t, an n by 1 vector, is Z_t minus μ, where Z_t is a vector of economic time-series (in our example, M, V^*, P and q) and μ is the vector of the means of Z.[9] A is an n by n matrix and ϵ_t is a vector of independent random disturbances that are stationary around zero–that is, $E(\epsilon_t) = 0$ and $E(\epsilon_t \epsilon_t') = \Omega$ for all t.

The possibility of k cointegrating vectors means there exists a k by n matrix β', of rank k, such that $\beta' Y_t$ is stationary in the sense that it is mean reverting. It is assumed that all of the elements of Y_t are integrated of the same order, $I(1)$; however, for the sake of illustration (here and elsewhere) we first consider the possibility that the elements of Y_t are $I(0)$. In this case, the long-run stationary solution to equation (4) is $Y_t = (I - A)^{-1}\epsilon_t$.[10] But there is no need to ask whether these variables are cointegrated, because clearly any linear combination of Y_t is mean reverting.

Now consider the case where each element of Y_t is $I(1)$. Assume further that the elements of Y_t are mutually independent so

9. Note that the system given by Y could be thought of as a multivariate p^{th}–order AR system that has been rewritten as an $AR(1)$ system.

10. $(1 - A)^{-1}$ exists if all of the eigenvalues of A are less than one in absolute value.

that $A = I$. In this case, no long-run equilibrium exists because the matrix $(I - A)$ is of rank zero. Since any linear combination of these independent $I(1)$ series must itself be $I(1)$, these variables are not cointegrated. No stationary linear combination of these variables exists.

Finally, consider the case where all of the univariate series are $I(1)$, but A is not an identity matrix. In this case, not every linear combination of Y_t is stationary because $(I - A)$ is not of full rank, that is, $(I - A)^{-1}$ does not exist; see fn.10. However, as we will show, some linear combinations of Y_t may be stationary. The number of such cointegrating vectors is determined by the rank of $(I - A)$. From a purely statistical point of view, cointegration places some restrictions on the matrix A. From an economic perspective note that economic theory determines the matrix A and therefore places some restrictions on the long-run behaviour of Y_t.

From a somewhat broader perspective then, the objective of cointegration analysis is to find an n by n matrix B', of rank n; such that $B'Y_t$ decomposes Y_t into its stationary and nonstationary components. This is accomplished by obtaining a k by n sub-matrix of B', β', of rank k such that the transformed series $\beta' Y_t$ is stationary. The k rows of B' associated with these stationary series are called "cointegrating vectors". The remaining $n - k$ unit root combinations are termed "common trends."[11]

2.2.3 Tests for Cointegration and Their Relation to Unit Root Tests

In illustrating tests for cointegration, we draw the analogy between tests for cointegration and tests for unit roots. Any autoregressive time-series of order p can be written in terms of its first difference, one lag level and $(p - 1)$ lag differences. Consider first the univariate case,

$$y_t^\star = a_1 y_{t-1}^\star + a_2 y_{t-2}^\star + \ldots + a_p y_{t-p}^\star + e_t, \tag{5}$$

11. This terminology stems from the fact that the original series are retrieved from the transformed ones by multiplying the transformed vector by the inverse of B. In general, this spreads the unit root "trend" processes through all the original series depending on where the zeros occur in B^{-1}. See Stock and Watson (1988).

where $y_t^\star = y_t - \mu$. The equation can be reparameterized as

$$\Delta y_t^\star = b_1 \Delta y_{t-1}^\star + b_2 \Delta y_{t-2}^\star + \ldots + b_{p-1} \Delta y_{t-p+1}^\star - c y_{t-p} + e_t, \quad (6)$$

where $b_i = -1 + a_1 + a_2 + \ldots + a_i$ and $c = 1 - a_1 - a_2 - \ldots - a_p$. Alternatively, equation (5) can be reparameterized as

$$\Delta y_t^\star = d_1 \Delta y_{t-1}^\star + d_2 \Delta y_{t-2}^\star + \ldots + d_{p-1} \Delta y_{t-p+1}^\star - c y_{t-1} + e_t, \quad (7)$$

where $d_{p-1} = -a_p$, $d_{p-2} = -a_p - a_{p-1}, \ldots, d_1 = -a_p - a_{p-1} - \ldots - a_2$.

The Dickey and Fuller test uses the t-statistic from the ordinary least squares regression to test the null hypothesis that the coefficient c in equation (7) is equal to zero. The t-statistic is the likelihood ratio test of the null hypothesis of a unit root, where the likelihood is conditional upon an initial value, y_0.[12]

Consider now the multivariate analogue to equation (5):

$$Y_t = A_1 Y_{t-1} + A_2 Y_{t-2} + A_3 Y_{t-3} + \ldots + A_p Y_{t-p} + \epsilon_t, \quad (8)$$

where $A_1, A_2, \ldots, A_p$ are $n \times n$ matrices. Equation (8), too, can be reparameterized as either

$$\Delta Y_t = \Gamma_1 \Delta Y_{t-1} + \Gamma_2 \Delta Y_{t-2} + \ldots + \Gamma_{p-1} \Delta Y_{t-p+1} - \psi Y_{t-p} + \epsilon_t, \quad (9)$$

or

$$\Delta Y_t = \theta_1 \Delta Y_{t-1} + \theta_2 \Delta Y_{t-2} + \ldots + \theta_{p-1} \Delta Y_{t-p+1} - \psi Y_{t-1} + \epsilon_t. \quad (10)$$

If the matrix $\psi = (I - A_1 - A_2 - \ldots - A_p)$ is full rank, then any linear combination of Y_t will be stationary. If ψ is a matrix of zeros, then any linear combination of Y_t will be a unit root process and, hence, nonstationary.[13]

This leaves an intermediate case where ψ is not a matrix of zeros, but is less than full rank. The rank of ψ, k, is the number of linearly independent and stationary linear combinations of Y_t that can be found. In other words, it is the number of linearly

---

12. An initialization assumption is necessary for unit root processes. The critical values of the test statistics change depending on whether μ is equal to zero or whether a time trend is included. Dickey and Fuller (1979) present versions of both tests where μ is assumed to be 0 and where μ is replaced with a linear time trend.

13. For example, let $p = 2$, if $A_2 = 0$ and $A_1 = I$, then $(I - A_1 - A_2) = 0$. In this case, all the elements of X_t are unit root processes and $(I - A_1 - A_2)$ is rank zero. But, if $A_2 = 0$ and A_1 is $(\varrho_1 \varrho_2 \cdots \varrho_n)$, where $|\varrho_1| < 1$ for all i, $i = 1, 2, \ldots, n$, then $(I - A_1 - A_2)$ is full rank and all the elements of X_t are stationary $AR(1)$ processes.

independent cointegrating relations among the variables in Y_t. Of course, the estimate of $\psi, \widehat{\psi}$, will almost always be of full rank in a numerical sense. The objective of tests for cointegration is to test for the rank of ψ by testing whether the eigenvalues of $\widehat{\psi}$ are significantly different from zero.[14] From this perspective, the tests for cointegration used by Stock and Watson (1988) and Johansen (1988) are multivariate analogues of the Dickey-Fuller test.[15]

2.3 IS THERE AN ECONOMIC INTERPRETATION OF COINTEGRATION VECTORS?

It should be clear now that, if several $I(1)$ variables are cointegrated, then one or more linear combinations of them will have a finite variance. But the broader question of the economic interpretation of such cointegrating vectors remains. In general, cointegrating vectors are obtained from the reduced form of a system where all of the variables are assumed to be jointly endogenous. Consequently, they cannot be interpreted as representing structural equations because, in general, there is no way to go from the reduced form back to the structure.[16]

---

14. To our knowledge Theil and Boot (1962) were the first to prepare a test for the statistical significance of the eigenvalues of $\widehat{\psi}$. In so doing, they were the first to develop a test for cointegration.

15. There are three versions of the univariate test for unit roots (zero mean, constant, non- zero mean and a constant linear time trend). Johansen (1988) presents the multivariate analogue of the likelihood ratio test when μ is zero. Johansen and Juselius (1990) extend this to the case where μ is non-zero. In practice, the difference between Y_t and Y_1, the first sample observation, should be used if the $\mu = 0$ assumption is maintained; hence, it seems reasonable to use the unconstrained μ versions of the tests; see Dickey, Bell and Miller (1986) for a discussion of which test to use. Note that the formulas for computing the test statistics are straightforward least-squares formulas. It is the distribution of the test statistics that are nonstandard.

16. Of course, because equation (7) is essentially a multivariate *VAR*, in principle, it should be possible to give these a "structural" interpretation by imposing identifying restrictions on the reduced-form parameters, as has been done recently for *VAR* models. For example, see Bernanke (1986) and Blanchard and Quah (1989).

Nevertheless, they might be thought of as arising from a constraint that an economic structure imposes on the long-run relationship among the jointly endogenous variables. For example, economic theory suggests that arbitrage will keep nominal interest rates, especially those on assets with the same or similar maturity, from getting too far away from each other. Thus, it is not surprising that such interest rates are cointegrated.[17]

2.3.1 Cointegration with Exogenous Variables

The importance of cointegration in economics can be highlighted by noting the close correspondence between cointegration and the typical linear, dynamic, simultaneous equation model used in macroeconomics and econometrics

$$AY_t = BY_{t-1} + CX_t + u_t. \qquad (11)$$

Y_t is a n by 1 vector of endogenous variables, X_t is a g by 1 vector of exogenous variables and u_t is a n by 1 vector of random disturbances, and A, B and C are matrices of unknown parameters. It is assumed that A^{-1} exists, so that the dynamic reduced form can be written as

$$Y_t = \pi Y_{t-1} + \Gamma X_t + \epsilon_t \qquad (12)$$

where $\pi(= A^{-1}B)$ and $\Gamma(= A^{-1}C)$ are matrices of unknown reduced-form parameters. Equation (12) contains only predetermined variables on the right-hand side, so that the dynamic response of Y_t to X_t can be studied by recursive substitution using equation (12). Letting L denote the lag operator ($LY_t = Y_{t-1}$), the result is written as

$$Y_t = (I + \pi L + \pi^2 L^2 + \ldots)\Gamma X_t + (I + \pi L + \pi^2 L^2 + \ldots)\epsilon_t. \qquad (13)$$

The infinite series $(I + \pi L + \pi^2 L^2 \ldots)$, evaluated at $L = 1$, converges to $(I - \pi)^{-1}$ if all of the eigenvalues of π are less than 1 in absolute value.

The expected value of equation (13), conditional on X_t, is

$$E(Y_t) = (I + \pi L + \pi^2 L^2 \ldots)\Gamma X_t$$
$$= (I - \pi L)^{-1}\Gamma X_t. \qquad (14)$$

17. Stock and Watson (1988) find that the nominal Federal funds, the three-month Treasury bill and one-year Treasury bill rates are cointegrated.

Equation (14) is used to investigate the long-run response of Y_t to a change in one or more of the elements of X_t. Assuming that X_t is a vector of zeros for all $t \leq 0$ and $X_t = \delta$ for all $t > 0$, and that the system converges, then, in the limit

$$E(Y) = (I - \pi)^{-1}\Gamma\delta. \tag{15}$$

Thus, $(I-\pi)^{-1}\Gamma\delta$ gives the long-run response of Y_t to a permanent change, δ, in X. Theil and Boot (1962) have termed $(I - \pi)^{-1}\Gamma$ the final-form multipliers.

The cointegration techniques of Johansen and Stock and Watson start with an equation similar to equation (12). There are two key differences, however. First, all of the variables are explicitly endogenous. Second, because π has unit eigenvalues corresponding to the common trends in the system, $(I - \pi)$ is not invertible, and therefore, the final-form multipliers in (15) are undefined.[18]

A related technical point is that the initial conditions are not transient. To see this in equation (12), let $X_t = 0$ for all t and substitute recursively to obtain

$$Y_t = \epsilon_t + \pi\epsilon_{t-1} + \pi^2\epsilon_{t-2} + \ldots + \pi^{t-1}\epsilon_1 + \pi^t Y_0. \tag{16}$$

Because the matrix π^i does not converge to a matrix of zeros as t approaches infinity, the initial condition is not transient and must be specified. Often, the initial vector, Y_0, is assumed to be a vector of zeros.

Can cointegration be used to determine the long-run response to changes in exogenous variables? In general, the answer is yes. However, how this is done, the interpretation of the estimated cointegrating vectors and the method of estimation, all depend on the assumptions made about X_t. Consider, for example, the case where the elements of X_t are non-stochastic and fixed in repeated samples. Rewrite equation (12) as

$$\Delta Y_t = -(I - \pi)Y_{t-1} + \Gamma X_t + \epsilon_t. \tag{17}$$

If Y_t is cointegrated, the n by n matrix $(I - \pi)$ is rank $k < n$, and π can be represented as $\alpha\beta'$, where α and β are n by k matrices

18. Finding unit eigenvalues in π is equivalent to finding zero eigenvalues in ψ. This is so because $\psi = I - \pi$ so that $|\psi - \lambda I| = |\lambda(I - \pi) - I| = |\pi - (1 - \lambda)I| = |\pi - \delta I| = 0$, so $\lambda = 0$ is equivalent to $\delta = 1$.

of rank k. Then

$$\beta' Y_t = (I - \beta'\alpha)\beta' Y_{t-1} + \beta'\Gamma X_t + \beta'\epsilon_t \tag{18}$$

is stationary.[19]

Equation (18) can be expressed as the k–dimensional system, with $\beta' Y_t = Y_t^\star$ i.e.

$$Y_t^\star = \pi^\star Y_{t-1}^\star + \Gamma^\star X_t + \epsilon_t^\star. \tag{19}$$

Because the k by k matrix $I - \pi^\star$ is full rank, $(I - \pi^\star)^{-1}$ exists. Thus, this lower-dimensional system has the steady-state representation,

$$Y_t^\star = (I - \pi^\star)^{-1}\Gamma^\star X_t + (I - \pi^\star)^{-1}\epsilon_t^\star. \tag{20}$$

The expected value of 20 gives the final-form multipliers for the cointegrated subsystem of Y_t. Consequently, there is a representation for the long-run average response of $\beta' Y_t$ to a change in X.

It must be remembered, however, that because the estimated cointegrating vector is conditional on the information set in the model, X_t needs to be included along with lagged differences of the endogenous variables when estimating the cointegrating vectors. Moreover, the distribution of the test statistic will not be invariant to X_t. Consequently, the Monte Carlo experiments used to derive the distributions of the test statistics used by, say, Johansen (1988), would have to be redone including X_t in the model.

In the case where X_t is stochastic, the method used to obtain estimated cointegrating vectors and their interpretation changes with the assumed structure of the model. A general form of such a model is

$$\begin{bmatrix} A & C \\ O & I \end{bmatrix} \begin{bmatrix} Y_t \\ X_t \end{bmatrix} = \begin{bmatrix} B & O \\ O & D \end{bmatrix} \begin{bmatrix} Y_{t-1} \\ X_{t-1} \end{bmatrix} + \begin{bmatrix} \epsilon_t \\ \eta_t \end{bmatrix}, \tag{21}$$

where it is assumed that

$$E(\epsilon_t \eta_t)' = \begin{bmatrix} \sigma_\epsilon^2 \Sigma_\epsilon & 0 \\ 0 & \sigma_\eta^2 \Sigma_\eta \end{bmatrix}$$

for all t. Again, assume that A^{-1} exists, so equation (21) can be

19. More technically, the eigenvalues of $I - \beta'\alpha$ are less than one in absolute value.

rewritten as

$$\begin{bmatrix} Y_t \\ X_t \end{bmatrix} = \begin{bmatrix} A^{-1}B & -A^{-1}CD \\ O & D \end{bmatrix} \begin{bmatrix} Y_{t-1} \\ X_{t-1} \end{bmatrix} + \begin{bmatrix} \xi_t \\ \eta_t \end{bmatrix} \tag{22}$$

or

$$\begin{bmatrix} Y_t \\ X_t \end{bmatrix} = \begin{bmatrix} \pi & \Gamma D \\ O & D \end{bmatrix} \begin{bmatrix} Y_{t-1} \\ X_{t-1} \end{bmatrix} + \begin{bmatrix} \xi_t \\ \eta_t \end{bmatrix}. \tag{23}$$

Cointegrating vectors and their interpretation depend on the assumptions made about π and D. For example, if all of the eigenvalues of π are less than one, so that conditional on X_t, Y_t is stationary, the common trends in the system are the result of X_t being non-stationary. The number of common trends will be equal to the number of unit roots in D. If $D = I$, for example, then there will be g common trends and n cointegrating vectors. In this case, there is a steady-state representation for the endogenous variables conditional on the exogenous variables.[20]

That is, the final-form multipliers of Theil and Boot are conditional on the exogenous variables. They do not exist, however, because the variance of the exogenous variables is unbounded and, hence, so is the unconditional variance of the endogenous variables.

If π also has some unit eigenvalues, however, then part of the nonstationarity of the system is due to instability in the dynamics of the endogenous variables in the system.[21]

In this case, the conditional final-form multipliers, analogous to those obtained by Theil and Boot, exist along the stationary directions given by the cointegrating vectors as before. An important aspect of this formulation is that, given the proposed structure of the system, the researcher can identify whether the common trends stem from the structural dynamics or are simply a manifestation of the stochastic properties of the exogenous variables. Under the assumption on $(\epsilon_t \eta_t)'$, estimating cointegrat-

---

20. A form of this model has been proposed recently by Hoffman and Rasche (1990). This is the Stock-Watson formulation, except that X_t is a set of latent variables. Conditional on the unobserved X_t, the endogenous variables are assumed to be stationary.

21. This so because

$$\begin{vmatrix} \pi - \lambda I & \Gamma D \\ O & D - \lambda I \end{vmatrix} = |\pi - \lambda I|\ |D - \lambda I| = 0.$$

ing vectors for the system given by the equations in (23) is the same as estimating them for the system given by the equations in (8).[22]

Hence, any of the multivariate methods discussed in the next section can be employed. The reader is cautioned, however, that these procedures would have to be modified to impose the upper triangular structure of the system given by the equation in (23).[23]

2.3.2 Should there be Many or Few Cointegrating Vectors?

Since it is possible to have $n-1$ cointegrating vectors in the system given by equation (17), the question naturally arises, is it better to have many or few cointegrating vectors? Providing a general answer to this question is difficult. Cointegrating vectors can be thought of as representing constraints that an economic system imposes on the movement of the variables in the system in the long-run. Consequently, the more cointegrating vectors there are, the "more stable" the system. Other things being the same, it is desirable for an economic system to be stationary in as many directions as possible.

To see why, consider a model with no common trends, so the system is stationary: Y_t never wanders "too far" from its steady-state equilibrium value (in the model discussed in the text, the vector of means). If there is one common trend and $n-1$ cointegrating vectors, however, $n-1$ of the variables must be solved for in terms of the n variables, and the structure of these variables follows a single common trend. Hence, there are only $n-1$ directions (as opposed to n in the previous example) where the variance is finite and one direction in which it is infinite. On the other hand, if there is only one cointegrating vector, the nth variable must be solved for in terms of the other $n-1$ variables. The system can wander off in $n-1$ independent directions; it is stable in only one direction.[24]

---

22. If these error terms are not independently distributed obtaining consistent estimates of the cointegrating vector will be more difficult. For an example of this case, see Stock and Watson (1989).

23. Indeed, both Stock and Watson (1988) and Johansen (1988) allow for the possibility of exogenous variables in the sense of equation (22); however, they do not impose the exogeneity restriction *ex ante*.

24. In terms of the model given by the equations in (22), when $(1-\pi)$

To see this point from a geometric perspective, consider the case were there are three endogenous variables that span R^3. If these variables are stationary, the system converges to a steady-state equilibrium, a point in R^3, and variation around that point is finite. If the variables are non-stationary, and there is one common trend, however, the system converges to a long-run equilibrium represented by a line, determined by the intersection of the planes defined by the two cointegrating vectors, in R^3.[25] This is a stationary equilibrium in the sense that the variance about this line is finite. If there are two common trends and one cointegrating vector, the long-run equilibrium is represented by a plane defined by the single cointegrating vector. The variables are unbounded in the plane, but cannot move too far from it. That is to say, that the variance in the plane is infinite, but the variance about the plane is finite. If there are no cointegrating vectors, the variables are free to wander anywhere in R^3–they are unbounded!

Consequently, when non-stationary variables are cointegrated, there exists a direction where a meaningful long-run relationship among them exists. The fewer the number of cointegrating vectors, the less constrained is the long-run relationship. Hence, all other things being the same, it seems desirable to have many cointegrating vectors.[26] Alternatively stated, we prefer economic models that have unique steady-state equilibria. Accordingly, researchers are interested not only in testing to see whether variables are cointegrated, but in obtaining precise estimates of the cointegrating vectors.

---

is full rank, the more cointegrating vectors in D, the more directions in which the "final-form" multipliers will exist.

25. Of course, the subsystem defined by equation (20) converges to a point in R^2. This point is given by the intersection of this plane and the line given by the intersection of the two cointegrating vectors.

26. Having said that, we should hasten to add that it is very unlikely that these tests will indicate that there are a large number of cointegrating vectors. These tests lack the power to reject the null hypothesis of zero cointegrating vectors. At a more practical level, it is well known that macroeconomic time-series data are highly correlated so that, typically, the generalized variance of the matrix of such variables is concentrated on relatively few principal components. See footnote 27.

2.4 ALTERNATIVE TESTS FOR COINTEGRATION

With a number of tests for cointegration being available, it is important to understand their similarities and their differences. The purpose of this section is to discuss the salient features of alternative tests for cointegration. Step-by-step instructions on how to perform two of the more difficult of these (the Stock-Watson and Johansen test) are presented in the Appendix.

The relationship among tests for cointegration developed by Johansen (1988), Engle and Granger (1987), Stock and Watson (1988) and Fountis and Dickey (1989) can be illustrated by considering the multivariate model

$$Y_t = A_1 Y_{t-1} + A_2 Y_{t-2} + \ldots + A_p Y_{t-p} + \epsilon_t. \tag{24}$$

Johansen reparameterizes equation (24) as

$$\Delta Y_t = \Gamma_1 \Delta Y_{t-1} + \Gamma_2 \Delta Y_{t-2} + \ldots + \Gamma_{p-1} \Delta Y_{t-p+1} - \psi Y_{t-p} + \epsilon_t, \tag{25}$$

where, as before, $\psi = (I - A_1 - A_2 \ldots - A_p)$. He then makes use of the fact that any n by n matrix ψ, of rank $k < n$ can be written as the product of two n by k matrices of rank k–that is, $\psi = \alpha\beta'$ where α and β are n by k matrices of rank k. He maximizes the likelihood function for Y_t conditional on any given β using standard least squares formulas for regression of ΔY_t on $\Delta Y_{t-1}, \Delta Y_{t-2}, \ldots, \Delta Y_{t-p+1}$ and $\beta' Y_{t-p}$. The solution to this maximization problem gives estimates of $\Gamma_1, \Gamma_2, \ldots, \Gamma_{p-1}$ and α conditional on β. Once this is done, β, or more specifically, the row space of β, is estimated.

In cointegration, it is only possible to determine the rank of $\alpha\beta'$.[27] Specifically, obtaining unique elements of α and β without imposing arbitrary constraints is impossible. The rank of ψ can be obtained by computing canonical correlations between ΔY_t and Y_{t-p}, adjusting for all intervening lags. Johansen chooses to put the lag level at the largest lag but, as has been shown earlier, this is not crucial.

In the Johansen approach, $-\alpha\beta'$ is the coefficient matrix on the lagged level. Upon pre-multiplying equation (25) by β', the last term in equation (25) is $\beta'\alpha\beta' Y_{t-p}$. Note that $\beta'\alpha$ has no zero eigenvalues so that $\beta' Y_t$ is a stationary vector time-series of dimension k. Thus, rows of β' are the cointegrating vectors.

---

27. The space spanned by the columns of the α and β' matrices is important because $\alpha\beta'$ can be obtained by many choices of matrices. To see this, note that for any nonsingular k by k matrix, H, $(\alpha H^{-1})(H\beta') = \alpha\beta'$.

Now consider adding to the k rows of $\beta', n-k$ more rows orthogonal to the columns of α. Denote the resulting matrix B'. Note that $B'\alpha$ is simply $n-k$ rows of zeros appended to the bottom of $\beta'\alpha$. In equation (25), after this transformation, the last $n-k$ rows involve only differences. Hence, $C_t = B'Y_t$ is a column vector of k stationary processes followed by $n-k$ unit root processes (common trends). In Johansen's approach, then, the matrix B' has the k cointegrating vectors as its first k rows and coefficients yielding the $n-k$ common trends as its last $n-k$ rows.

By standard results in unit root estimation, the vector $C_t = B'Y_t$ is such that $N^{-2}\Sigma C_t C_t'$ (where N is the series' length) converges to an n by matrix with zeros everywhere except in the lower right $n-k$ by $n-k$ submatrix. This result underlies the other approaches to cointegration.

For example, Stock and Watson point out that if $W_t = (B')^{-1}C_t$, then the sum of squares and cross products matrix $N^{-2}\Sigma W_t W_t'$ converges to the same limit as $(B')^{-1}\Sigma C_t C_t' B^{-1}$. This limit matrix has rank $n-k$ and, thus, has $n-k$ nonzero eigenvalues. Because the generalized variance of a sum-of-squares and cross-products matrix is equal to the sum of the eigenvalues of this matrix they suggest a test based on computing the largest eigenvalues of the sum-of-squares and cross-products matrix.[28]

In multivariate statistics, a vector random variable Y can be transformed into a canonical vector random variable $C = TY$ by choosing T (a matrix of constants) so that the elements of C are uncorrelated. The elements of C are referred to as "principal components" of Y. Stock and Watson reason that if a vector process has $n-k$ common trends (that is, the vector $C_t = B'Y_t$ has $n-k$ unit root processes and k stationary ones), then the $n-k$ principal components with largest variance should correspond

28. For example, let X be an n by k matrix and let $J = (X'X)$ be a k matrix of rank k. Then there exists a k by k matrix T such that $T'T = \Lambda$ is a diagonal matrix with the eigenvalues of J on the diagonal. The columns of T are the eigenvectors of J. These eigenvectors are called "principal components" because the generalized variance of $J, |J|$, is equal to the trace of Λ. Because often many of the eigenvalues of J are close to zero, its generalized variance can be closely approximated by a relatively few "principal components." A good discussion of this can be found in Dhrymes (1970, pp. 53-59).

to the unit root processes or "common trends". This reasoning is based on the previously stated notion that the normalized sum-of-squares and cross-product matrix $N^{-2}\Sigma Y_t Y_t'$ converges to a singular limit so the variances of the $n-k$ common trends correspond to principal components giving the nonzero eigenvalues of the limit matrix. The k principal components with smaller variances correspond to the zero eigenvalues of the limit matrix and each of the k rows is a cointegrating vector.

2.4.1 A Note about Distributions

Having more than one lag in the model introduces matrices of nuisance parameters (that is, parameters which must be estimated but are not of primary interest) that affect the form of the test statistics. How the existence of these parameters complicates the test procedures is illustrated for the univariate case.[29] Consider a simple version of the univariate model given by equation (5):

$$(y_t - \mu) = (\varrho + \lambda)\ (y_{t-1} - \mu) - \lambda\varrho(y_{t-2} - \mu) + e_t. \tag{26}$$

If $\varrho = 1$, μ drops out. the series y_t is a unit root process with no tendency to move toward any level. In this case, however, the test of the null hypothesis that $\varrho = 1$ is complicated by the presence of a nuisance parameter, λ. Regardless of ϱ, as noted before, the model can be reparameterized as

$$\Delta y_t = -(\varrho - 1)(\lambda_1)(y_{t-1} - \mu) + \lambda\varrho\Delta y_{t-1} + e_t. \tag{27}$$

Notice that the coefficient on $(y_{t-1} - \mu)$ is now a *multiple* of $\varrho - 1$. In ordinary regression, multiplying a regressor by a constant changes the distribution of the regression coefficient by a multiplicative constant, but does not change the distribution of its t-statistic. The same holds true asymptotically in cointegration.[30]

---

29. These ideas can also be found in Fuller's text (1976) or in Dickey, Bell and Miller (1986).

30. Johansen, using a multivariate analogue of the Dickey-Fuller τ statistic, only needs to have a correctly specified likelihood function. Stock and Watson, using a multivariate analogue of the normalized coefficient tests $n(\widehat{\varrho} - 1)$, $n(\widehat{\varrho}_\mu - 1)$, or $n(\widehat{\varrho}_\tau - 1)$ of Dickey and Fuller need to adjust for this fact multiplier (corresponding to $1 - \lambda$ in our univariate example).

If λ is known, the estimated coefficient on $(y_{t-1} - \mu)$ could be divided by $(1-\lambda)$ and the limit distribution $n(\widehat{\varrho}_\mu - 1)$ listed in Fuller (1976, section 8.5) could be obtained. If λ is unknown, however, an approximation must be obtained by dividing by $(1-\widehat{\lambda})$, where $\widehat{\lambda}$ is estimated by regressing Δy_t on Δy_{t-1}, suggested by imposing the null hypothesis $\varrho = 1$.

Alternatively, the univariate model could be written as

$$\Delta(y_t - \lambda y_{t-1}) = (\varrho - 1)(y_{t-1} - \lambda y_{t-2} - \mu(1-\lambda)) + e_t. \tag{28}$$

Since $\widehat{\lambda}$ is a consistent estimate of λ, it can be used to filter y_t. That is, $\widehat{F}_t = y_t - \widehat{\lambda} y_{t-1}$. $\Delta\widehat{F}_t$, can then be regressed on $\Delta\widehat{F}_{t-1}$ (with an intercept) or on $F_{t-1} - \bar{F}$, where $\bar{F}$ is the mean of the series, $\widehat{F}_t$, to get a test statistic asymptotically equivalent to $n(\widehat{\varrho}_\mu - 1)$.

2.4.2 Other Approaches to Cointegration

The approach of Fountis and Dickey (1989) is similar to that of Stock and Watson but only allows for the possibility that there is one unit root. As such, it is much less general than either of the above approaches. It does, however, provide estimates of the cointegrating vectors and common trend, based on the coefficient matrix of the lagged levels. Although there is an asymptotic link between variances and lag coefficients as established above, the actual definition of a unit root process is in terms of lag relationships. All approaches to cointegration use lags in *testing*. However, in *transforming* Y_t, Stock and Watson use only the variance-covariance matrix while Fountis and Dickey (and Johansen) use the lag information.

Finally, we mention an approach by Engle and Granger which is especially easy to use and to explain in the bivariate case. Consider two univariate series y_t and z_t. The first step is to check that each is a unit root process. Next regress y_t on z_t (or z_t on y_t) getting the linear combination $y_t - \hat{b}z_t$ with the smallest variance. Notice that this does not use lag information to obtain the transformation $y_t - \hat{b}z_t$. The next step is to test if the series $y_t - \hat{b}z_t$ is stationary (if so, the vector $(1,\ \hat{b})$ is a cointegrating vector). Had the parameter b been pre-specified instead of estimated, an ordinary Dickey-Fuller test would be appropriate. Since the data are used to estimate b, however, the previously mentioned problems analogous to the "multiple comparisons" problem exist. Engle and Granger have provided the appropriately adjusted percentiles. Other than using these special tables, this approach uses only the standard univariate unit root testing strategy.

One problem with this approach is that it requires the researcher to choose one of the jointly endogenous variables to put on the left-hand side. While the test is asymptotically invariant to this so-called direction normalization rule, the test results may be very sensitive to it in finite samples. Indeed, practical experience indicates that the result of the test depends qualitatively on which variable is chosen to be on the left-hand side. The alternative multivariate approaches have the advantage that all of the variables are explicitly endogenous, so the researcher does not have to make such arbitrary normalization choices.

2.5 AN APPLICATION OF COINTEGRATION: THE DEMAND FOR MONEY

One important macroeconomic relationship that has received considerable attention is the link between money and income. This relationship, embedded in the demand for money, is commonly represented by the income velocity of money. Since the income velocity of $M1$ has drifted upward over time, it does not appear to be stationary. Furthermore, formal tests indicate that the income velocity of $M1$ is not stationary, indicating that $M1$ and income, in the form of income velocity, are not cointegrated.[31]

In contrast, the income velocity of $M2$ appears to move around an unchanged mean, and formal tests suggest that $M2$ and income are cointegrated.[32]

These results have been interpreted by some as evidence against the existence of a stable long-run demand for $M1$ and

31. Strictly speaking, the nonstationarity indicates only that the cointegrating vector for $M1$ and nominal GNP is not $(1, -1)$. For example, see Nelson and Plosser (1982). Also, see Engle and Granger (1987) who test for cointegration between $M1$ and income.

32. For example, see Engle and Granger (1987). However, the results for $M2$ appear to be sensitive to the sample period and how the test is performed. See Hallman, Porter and Small (1989, 1990) and Hafer and Jansen (1991).

for the existence of a stable long-run demand for $M2$.[33]

But if $M1$ is the relevant measure of *money,* the Fisher equation suggests that there should be at least one cointegrating relationship between $M1$, its velocity, real income and the price level. The specification of tests for cointegration depend on the specification of the demand for money. Consequently, it is important to review the theory of money demand before performing tests for cointegration.

A general specification for the long-run demand for money is

$$M^d = f(P,\ Q,\ Z) \tag{29}$$

where M and Q denote the nominal money stock and the nominal income level, respectively, P denotes the level of prices and Z denotes all other variables that affect money demand–for example, current and expected future real interest rates, the expected rate of inflation, etc. Assuming that economic agents do not suffer from money illusion, equation (29) can be written as

$$M^d/P = m^d = f(Q/P,\ Z) = f(q,\ Z). \tag{30}$$

That is, the demand for real money balances, m^d, is a function of real income q, and some other variables. Furthermore, it is commonly assumed that the demand for money is homogeneous of degree one in real income, so that equation (30) can be written as

$$m^d/q = h(Z), \tag{31}$$

where $h(Z)$ is the famous k in the Cambridge cash balance equation which is the reciprocal of the income velocity of money. In equilibrium, the demand for real money equals the supply of real money, m^s, so that $h(Z)$ is observed simply as the ratio of real money stock to real income.

2.5.1 The Velocity of M1 and M2

Consider now two alternative monetary aggregates, $M1$ and $M2$. In this framework, the reciprocals of their velocities can be written, respectively, as

$$m1/q = h1(Z), \quad \text{and} \tag{32}$$

33. For example, Hallman, Porter and Small (1989) have predicted their P–star model on such a long-run stable demand for $M2$.

$$m2/q = h2(W). \quad (33)$$

The specification for $M2$ allows for the possibility that there are factors that affect its demand that do not affect the demand for $M1$–that is, Z is a subset of W. Since $M2$ is simply $M1$ plus some other financial assets, equation (33) can be written as

$$m1/q + nm1m2/q = h1(Z) + v(W) \quad (34)$$

where $nm1m2$ denotes the real non-$M1$ components of $M2$ and $v(W) = [h2(W) - h1(Z)]$, hereafter, called the reciprocal of $NM1M2$ velocity.[34]

The above analysis has two important implications. First, because velocity is not directly and independently observable, its proxy must be specified to perform tests for cointegration. Second, if in fact, $M1$ and income are not cointegrated but $M2$ and income are, then a long-run stable inverse relationship must exist between $h1(Z)$ and $v(W)$. On average, movements in $h1(Z)$ must be offset by movements in $v(W)$, that is, $h1(Z)$ and $v(W)$ must be cointegrated.

A simple analysis of $M1$, $M2$ and GNP data is consistent with this conjecture. Figures 2.1 and 2.2 show the observed income velocities of $M1$ and $M2$, that is, $(q/m1)$ and $(q/m2)$respectively, for the period from 1953:1 to 1988:4. $M1$ velocity trends upward through the early 1980s and then declines. In contrast, $M2$ velocity appears to cycle with no apparent trend.[35] Also, the sharp break in the pattern of $M1$ velocity in the early 1980s is not as apparent in $M2$ velocity.

Figure 2.3 shows the reciprocals of $M1$, $M2$ and $NM1M2$ velocities. As expected, the downward trend in the reciprocal of $M1$ velocity through 1980 appears to be matched by an upward trend in the reciprocal of the velocity of $MN1M2$.Also, a comparison of the series in Figure 2.3 reveals that much of the variability in the reciprocal of $M2$ velocity is associated with variability in

34. These relationships hold for seasonally adjusted $M2$ as well, because $M1$ and the non–$M1$ components of $M2$ are seasonally adjusted separately and added together.

35. Dickey-Fuller tests cannot reject the null hypothesis of a unit root over some sample periods. Figure 2.2 suggests that these results are likely driven by the low power of the test against the alternative of a large, but stationary, root. Hence, one might wish to rely on his eyes rather than the formal test results.

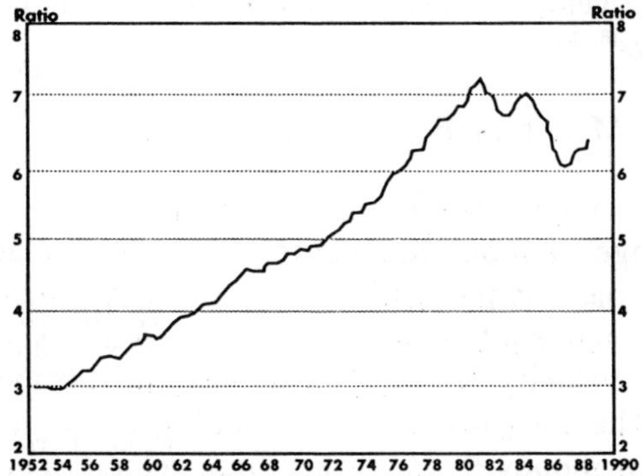

Figure 2.1 The income velocity of M1 1953:1 to 1988:4

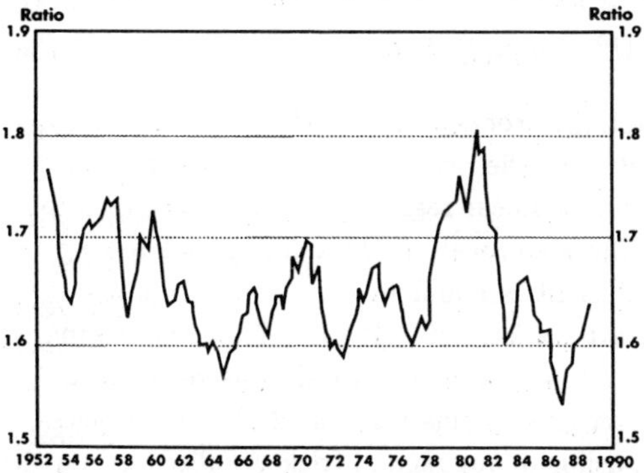

Figure 2.2 The income velocity of M2 1953:1 to 1988:4

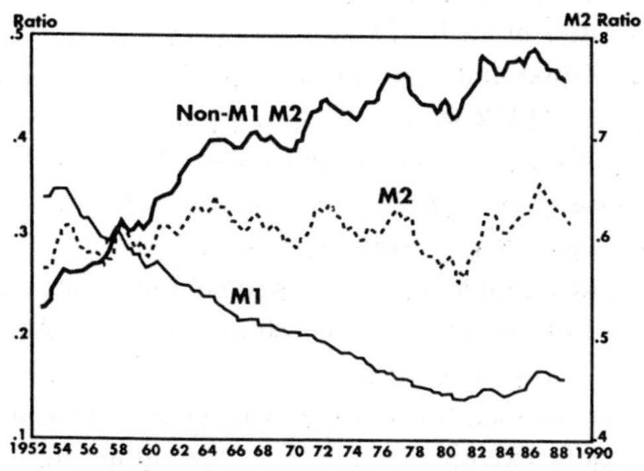

Figure 2.3 The reciprocals of income velocities of M1, M2 and the non–M1 components of M2 1953:1 to 1988:4

the reciprocal of $NM1M2$ velocity, rather than variability of the reciprocal of $M1$ velocity.[36, 37]

2.5.2 The Velocity of the Monetary Base

The concept of income velocity can be extended directly to a broader or narrower range of monetary aggregates. One such aggregate, the monetary base, is particularly important because the monetary authority can control it fairly well. While the observed monetary base velocity is simply the ratio of income to the monetary base, the measure is only meaningful in the context of the demand for money.[38] Monetary base velocity can be incorporated into the money demand framework by noting that the nominal money supply, M^S, can be expressed as

$$M^S = mm(H)MB \tag{35}$$

36. At a more formal level, the proposition that the decline in the reciprocal of $M1$ velocity is just offset by the rise in the reciprocal of the velocity of $NM1M2$ was tested by a simple linear regression in which the reciprocal of $NM1M2$ is regressed on the reciprocal of $M1$ velocity and testing the null hypothesis that the coefficient is equal to -1. The estimated slope coefficient was -1.055 with a t-statistic of -41.47. While the estimated coefficient was very close to -1, the null hypothesis was rejected at the 5% significance level. The t-statistic in this test was 2.17. Nevertheless, a formal test for cointegration, using an augmented Dickey-Fuller test, on the residuals from this equation suggests that these variables are cointegrated. This proposition also was tested by estimating a simple linear time trend for the reciprocals of the velocities of each $M1$ and $NM1M2$ and testing the hypothesis that trends are equal and opposite in sign. The estimated trend coefficients for $M1$ and $NM1M2$ were $-.00143$ and $.00154$, respectively, and both were statistically significant at well below the 5% level. But , again, despite the closeness of the estimates, the null hypothesis was rejected at the 5% significance level. The F-statistic was 8.67.

37. This can be illustrated by a simple linear regression of the reciprocal of $M2$ velocity on the reciprocals of the velocities of each $M1$ and $NM1M2$. These regressions indicate that the reciprocal of the velocity of $NM1M2$ explains 19% of the variation in the reciprocal of the velocity of $M2$, while the reciprocal of $M1$ velocity alone explains only .03% of the variation.

38. For example, it is not reasonable to obtain this result by simply assuming that the monetary base is the appropriate measure for money because it is composed of currency and bank reserves.

where MB denotes the adjusted monetary base and mm denotes the money multiplier which is a function of a set of variables H, for example, interest rates, portfolio preferences of the public, etc. Dividing both sides of equation (35) by the price level and substituting the result for m^d in equation (31) yields

$$mm(H)mb = qh(Z), \tag{36}$$

where mb denotes the real monetary base. Because the money multiplier is not observed independently, test of cointegration involving the monetary base must include not only q and Z, but also H.[39]

2.5.3 Empirical Results

The empirical worggr k is presented in two parts. The first presents tests for cointegration using methodologies suggested by Johansen, Stock-Watson and Engle-Granger. The results in this part are presented only for $M1$. In the second part, the analysis is extended to a broader set of monetary aggregates using the Johansen approach. The Johansen methodology was chosen because it is based on the well-accepted likelihood ratio principle. Moreover, recent Monte Carlo evidence by Gonzalo (1989) suggests that Johansen's maximum likelihood technique for estimating and testing cointegrating relationships performs better than both single equation methods and alternative multivariate methods. Nevertheless, because the other two approaches are widely known and the Engle-Granger approach is especially widely used in the empirical literature to date, we report test results for $M1$ using all three techniques.

In the second part, the analysis is extended to other monetary aggregates. These aggregates are the adjusted monetary base (calculated by the Federal Reserve Bank of St. Louis), $M2$ and the non-$M1$ components of $M2$, denoted $NM1M2$. When the monetary base is used, the ratio of currency to total check-

39. It is not necessary that H include variables that are not included in Z. If it does not, however, there is an identification problem. That is, one cannot tell the difference between the above model and simply treating the monetary base as the appropriate monetary aggregate. Since this possibility is difficult to conceive of, it is useful if H includes variables that are not in Z, as in our empirical work which follows.

able deposits, denoted K, is also included because it is the most important determinant of the money multiplier.[40]

In both parts, the income and price level measures are real *GNP*, q, and the *GNP* deflator, P, respectively. Two measures of nominal interest rates, R, are used: the three-month Treasury bill rate, *R3M*, and the yield on 10-year government securities, *R10Y*. The data consist of quarterly observations from 1953:2 to 1988:4 and all data are transformed to natural logarithms.

2.5.4 Tests for the Order of Integration

Before testing for cointegration, the order of integration of the individual time-series must be determined. Tests for unit roots are performed on all of the data using the augmented Dickey-Fuller test with three lagged differences. The null hypothesis is that the variable under investigation has a unit root, against the alternative that it does not. The substantially negative values of the reported test statistic lead to rejection of the null hypothesis.

The tests are performed sequentially. The first column in the top half of Table 2.1 reports tests of stationarity of the levels of the time-series about a non-zero mean. The critical values of the test statistic [t_μ] are tabulated in Fuller (1976) and discussed in Dickey and Fuller (1979). The reported test statistics indicate that the null hypothesis cannot be rejected for any variable. We then test for stationarity about a deterministic time trend, using the Dickey-Fuller statistic [t_τ]. The results of this test are given in the second column in the top half of Table 2.1. Critical values for this test statistic are tabulated in Fuller (1976). With the exception of *R3M*, the null hypothesis that the time-series has a unit root cannot be rejected.

The bottom half of Table 2.1 reports results for the augmented Dickey-Fuller test on first differences of the variables. The null hypothesis of a unit root is rejected for all of the time-series using differenced data. These results are broadly consistent with the hypothesis that the individual time-series are

40. A simple linear regression of the multiplier (or its growth rate) on K (or its growth rate) using quarterly data, indicates that K alone explains 95% of the variation in the level of the multiplier. Moreover, the growth rate of K alone explains 84% of the variation in the growth rate of the multiplier.

Table 2.1
Augmented Dickey-Fuller Test for a Unit Root

	t_μ	t_τ
Variables	*3 lags*	*3 lags*
$M2/P$	−0.801	−2.456
$M1/P$	−0.817	−1.444
MB/P	0.308	−2.285
q	−0.723	−2.167
$R3M$	−2.689	−3.884*
$R10Y$	−1.873	−2.444
K	−0.561	−2.274
$NM1M2/P$	−2.150	−1.580
$\Delta(M2/P)$	−4.006*	
$\Delta(M1/P)$	−3.618*	
$\Delta(MB/P)$	−3.067*	
Δq	−5.923*	
$\Delta R3M$	−6.634*	
$\Delta R10Y$	−5.586*	
ΔK	−5.127*	
$\Delta(NM1M2/P)$	−3.855*	

Notes:

* Indicate statistical significance at the 5% level.

Critical values $t_\mu(T = 100) = -2.89$, $t_\tau(T = 100) = -3.45$.

individually $I(1)$. Because these data appear to be stationary in first differences, no further tests are performed.

2.5.5 Tests for Cointegration Using Three Methodologies

Tests for cointegration for real $M1$, real income and either $R3M$ or $R10Y$ using methodologies proposed by Johansen, Stock-Watson and Engle-Granger are presented in Table 2.2. For the Johansen and Engle-Granger tests three lagged differences were used. Both the test statistics and the estimated cointegrating vector, setting the coefficient on $(M1/P)$ equal to one, are reported. The estimated cointegrating vector is reported even when the test does not indicate cointegration.

For the Johansen method, there are two test statistics for the number of cointegrating vectors: the trace and maximum eigenvalue statistics. In the trace test, the null hypothesis is that the number of cointegrating vectors is less than or equal to k, where k is 0, 1 or 2. In each case the null hypothesis is tested against the general alternative. The maximum eigenvalue test is similar, except that the alternative hypothesis is explicit. The null hypothesis $k = 0$ is tested against the alternative that $k = 1$, $k = 1$ against the alternative $k = 2$, etc. The critical values for these tests are tabulated by Johansen and Juselius (1990). For the trace test, the hypotheses $k \leq 1$ and $k \leq 2$ cannot be rejected for either of the two interest rates, while the hypothesis $k = 0$ can be rejected.[41]. Consequently we conclude that there is one cointegrating vector.

Turning to the maximum eigenvalue test, the hypothesis $k = 0$ is uniformly rejected in favour of the alternative $k = 1$. Consequently this test indicates that real $M1$ is cointegrated with real income and either of the two nominal interest rates. Moreover, there appears to be a single cointegrating vector. The maximum eigenvalue test of $k = 1$ vs. $k = 2$ fails to reject the null hypothesis of $k = 1$. Thus, there are two common trends and one cointegrating vector.

The Johansen test produced results that were markedly different from those obtained using either the Engle-Granger or Stock-Watson methodologies. Because the results of the Engle-Granger can change with the variable chosen as the dependent variable, the test was performed with each variable on the right-hand side. Not surprisingly, the test was sensitive to this choice. The test indicated cointegration only when $R3M$ was the dependent variable although the test indicated cointegration at slightly higher significance level when real GNP, that is q, is the dependent variable and $R3M$ is used as the interest rate variable. It is interesting to note, however, that the estimated cointegrating vectors obtained from the Johansen and Engle-Granger approaches are nearly identical when both indicate cointegration.

41. Johansen and Juselius (1990, p.9) pointed out that "One would, however, expect the power of this procedure [the trace test] to be low, since it does not use the information that the last three eigenvalues have been found not to differ significantly from zero. Thus one would expect the maximum eigenvalue test to produce more clear cut results."

Table 2.2
TESTS FOR COINTEGRATION FOR M1

Test Statistics					*Cointegrating Vector*		
					q	$R3M$	$R10Y$
			Johansen Test[1]				
Trace			*Max. Eigenvalue*				
$k=0$	$k \leq 1$	$k =\leq 2$	$k=0$	$k=1$			
32.3*	5.2	1.2	27.1*	4.0	.680	−.369	—
28.4	6.0	1.8	22.4*	4.2	.845	—	−.570
			Stock – Watson Test[2]				
$q(3,2)$							
−24.91					.398	−.152	—
−15.61					.635	—	−.353
			Engle – Granger Test[3,4]				
Dependent Variable	*Test Statistics*						
$\frac{M1}{P}$	−2.26				.353	−.131	—
	−2.08				.558	—	−.303
q	−3.92				.671	−.271	—
	−3.18				.794	—	−.449
R	−4.86*				.730	−.359	—
	−3.36				.826	—	−.496

Notes:

* Indicates statistical significance at the 5% level.

1. Critical values:

Trace			*Max. Eigenvalue*	
$k=0$	$k \leq 1$	$k \leq 2$	$k=0$	$k=1$
31.3	17.8	8.1	21.3	14.6

2. Critical values for $q(3,2)$ is −31.5.
3. Critical values for ADF taken from Engle and Yoo (1987)

100 *observations*	200 *observations*
−3.93	−3.78

4. For an interpretation of the coefficients in the last three columns see footnote 42.

This is not the case, however, when the Engle-Granger or Stock-Watson tests do not indicate cointegration.[42]

Although evidence concerning cointegration between real $M1$, interest rates and output is sensitive to the method used, the results using Johansen methodology are similar to those of Hoffman and Rasche (1989) using monthly data and the same methodology. Moreover, for the Johansen results, the hypothesis that the normalized coefficient on output is unity is not rejected using either interest rate, while the hypothesis of a zero coefficient for the interest rate is rejected for both interest rates (see Table 2.4).[43] Thus, the data appear to support the notion that there is a stable long-run relationship between real $M1$, real income and interest rates.

2.5.6 Cointegration Using Alternative Monetary Aggregates

Tests for cointegration using real $M2$ and real $NM1M2$ are presented in Table 2.3. Both the trace and maximum eigenvalue tests indicate one cointegrating vector for $M2$, and both indicate cointegrating vectors for $NM1M2$.[44]. The previous discussion of the relationship between $M1, M2$ and $NM1M2$ suggests that certain long-run relationships should exist between the estimated cointegrating vectors using the various aggregates. In order to examine this suggestion the estimated cointegrating vectors, normalized on the *real* value of the respective aggregate, are presented in Table 2.4, along with tests of the hypotheses that the coefficient on output is unity and the coefficient on the interest rate is zero. These test statistics are asymptotically distributed $\chi^2(1)$.

---

42. The estimated cointegrating vector from the Engle-Granger method, when the equation is normalized on real output, is $(M1/P) = -.271R + .671q$. While the estimated coefficient on the three-month Treasury bill rate is markedly different from that when cointegration is indicated, the estimated coefficient on output is nearly identical.

43. The $\chi^2(1)$ statistics for the test of the coefficient on output are 2.23 and .61, respectively, for the three-month and 10-year rates, and the test statistics for the interest rates are 21.37 and 14.66, for the two rates, respectively.

44. If a 10% significance level is used, the test indicates there are two cointegrating vectors for $M2$ when the 10-year bond rate is used.

Table 2.3
Tests for Cointegration for the Broader Monetary Aggregates

	Trace Statistic			*Max.Eigenvalue*	
Aggregate	$k = 0$	$k \leq 1$	$k \leq 2$	$k = 0$	$k = 1$
$M2$	33.7*	13.9	1.6	19.8	12.3
	38.7*	17.3	4.4	21.4*	12.9
$NM1M2$	53.4*	20.6*	2.7	32.8*	17.9*
	39.7*	16.3	3.7	23.4*	12.6

Notes:

* Indicate statistical significance at the 5% level.

Critical Values	*Trace*			*Max. Eigenvalue*	
	$k = 0$	$k \leq 1$	$k \leq 2$	$k = 0$	$k = 1$
	31.3	17.8	8.1	21.3	14.6

The apparent nonstationarity of $M1$ velocity and the stationarity of $M2$ velocity implies that $M1$ and $NM1M2$ must have compensating nonstationary behaviour. This suggests that the sum of the income elasticities or interest elasticities for $M1$ and $NM1M2$ should equal that of $M2$. While there are no formal tests of these cross-equation restrictions, the point estimates in Table 2.4 indicate that, with the exception of the income elasticity when *R3M* is used, these restrictions do not do too much violence to the data. For the three-month rate, the sum of the interest elasticities for $M1$ and $NM1M2$ is $-.15$ compared with the estimated elasticity for $M2$ of .02; the sum of the income elasticities is 1.72, compared with an estimated income elasticity for $M2$ of .97. For the 10-year rate, the sum of the elasticities is $-.08$, compared with the estimated elasticity of $-.03$; the sum of the income elasticities is .92, compared with an estimated income elasticity of 1.04. Nevertheless, the fact that the hypothesis that the income elasticity is unity cannot be rejected for either $M1$ or $M2$ is troubling.

2.5.7 Cointegration and the Monetary Base

As was noted earlier, the monetary base must be regarded as

Table 2.4
Normalized cointegrating Vectors and Hypothesis Tests

Aggregate	*Cointegrating Vector*			*Hypothesis Test*		
	q	$R3M$	$R10Y$	q^1	$R3M^2$	$R10Y^2$
$M1$	0.680	−0.369	—	2.23	21.37*	—
	0.845	—	−0.570	0.61	—	14.66*
$M2$	0.971	0.016	—	0.20	0.16	—
	1.044	—	−0.031	0.21	—	0.24
$NM1M2$	1.043	0.217	—	0.09	8.58*	—
	0.078	—	0.646	2.72	—	4.30*

Notes:

* indicates statistical significance at the 5% level.

1. Null hypothesis is that the income elasticity is one.
2. Null hypothesis is that the interest elasticity is zero.

a supply-side variable, and cointegration of the monetary base with income and interest rates arises due to the relationship between the monetary base and the relevant money stock measure. Consequently, it is necessary to include a proxy for the money multiplier in an investigation of cointegration for the monetary base. Because the primary determinant of the multiplier is the currency-deposit ratio, K, it is included along with the real monetary base, real income and an interest rate in tests for cointegration.

The results, presented in Table 2.5, indicate that there are two cointegrating vectors linking the real monetary base, real income, the nominal interest rate and K when $R3M$ is used, but only one cointegrating vector when $R10Y$ is used. Because a cointegrating vector merely represented a long- run, stable relationship among jointly endogenous variables, in general, they cannot be interpreted as structural equations. Consequently, neither of the estimated cointegrating vectors necessarily represents either the long-run demand for or long-run supply of money. All that can be said is that there are two linear combinations for which the variance is bounded. Nevertheless, it is interesting to note that the second reported cointegrating vector is broadly

Table 2.5
Tests for Cointegration Using the Monetary Base

Trace Test				*Max.Eigenvalue*			*Cointegrating vector*			
$k=0$	$k\leq 1$	$k\leq 2$	$k\leq 3$	$k=0$	$k=1$	$k=2$	q	K	$R3M$	$R10Y$
65.5*	30.6	7.4	2.3	34.9*	23.2*	5.1	.465	.333	−.271	—
52.6*	22.5	5.3	1.4	30.1*	17.2	3.9	1.239	−.540	—	−.269

Notes:

* Indicate statistical significance at the 5% level.

consistent with equation (36).[45]

Nevertheless, because a stable long-run demand for money implies that there is a stable long-run relationship between real money, real income and either a short- or long-term interest rate. Consequently, these results are consistent with the proposition that the long-run demand for money is stable, even though they may not be estimates of the long-run money demand function itself.[46]

They also suggest that the reason $M2$ velocity is stable is because it includes transactions and non-transactions components that are close substitutes for each other in the long run. In particular, the upward drift in $M1$ velocity appears to be due largely to a relatively steady shift from $M1$ to the non-transactions deposits in $M2$. The magnitude of the trend movements in these variables is approximately equal so that $M2$ velocity is essentially trendless over the estimation period.[47]

45. Taking the log of equation (36) and letting the money multiplier be a function of both K and the interest rate and $h(Z)$ be solely a function of the interest rate, results in $\ln\ mb = \ln\ q + \ln\ h(R) - \ln\ mm(R,K)$, where $h' < 0$ and $\partial mm/\partial R > 0$ and $\partial mm/\partial K < 0$.

This equation implies that the long-run elasticity of $\ln\ mb$ with respect to $\ln\ q$ is unity, the elasticity of $\ln mb$ with respect to R is negative, but, smaller than the estimate for the long-run demand for money, and that the elasticity with respect to K is negative.

46. These results are both quantitatively and quantitatively similar to those obtained by Hoffman and Rasche (1989).

47. The stable long-run relationship between real income, the real monetary base, nominal interest rates and the currency-deposit ratio, is

2.6 SUMMARY AND CONCLUSIONS

This paper reviews the concept of cointegration, notes the relationship between tests for it and common tests for unit roots and considers its implications for the relationship among real money balances, real income and nominal interest rates. We argue that if $M2$ and nominal income are cointegrated, while $M1$ and nominal income are not, there necessarily exists a stationary long-run relationship between $M1$ and the non-$M1$ components of $M2$. We also argue that, if $M1$, real income and the nominal interest rate are cointegrated, the same could be true for real income, the nominal interest rate, the monetary base and a proxy for the monetary base/money multiplier.

Tests for cointegration among real $M1$, real income and one of two interest rates using three alternative procedures show that the results are sensitive to the method used. Nevertheless, the technique proposed by Johansen indicates that there is a single cointegrating relationship among these variables. While the cointegrating vector cannot be interpreted as the long-run demand for money, the estimated long-run income and interest elasticities are consistent with those often hypothesized and estimated for the long-run demand for money.

We also show that the hypothesized long-run relationship for the cointegrating vectors for $M1, M2$ and the non-$M1$ components of $M2$, namely that the sum of the income and interest elasticities for $M1$ and the non-$M1$ components of $M2$ equal the income and interest elasticities of $M2$, is supported by the data. Finally, we show that if the currency-deposit ratio is used to proxy the monetary base multiplier, the real monetary base, real income, the interest rate and the currency-deposit ratio are cointegrated.

The last two results are consistent with the notion of a stable long-run relationship between monetary aggregates and prices when both real income and nominal interest rates are taken in to account. Moreover, since a stable long-run relationship between real money, real income and nominal interest rates seems to exist, it can be said that there is potential for achieving price level stability by controlling the growth rates of either $M1$ or the monetary base.

---

also consistent with the idea recently put forth by McCallum (1987) that nominal GNP can be controlled in the long-run by monetary base targeting.

Appendix to Chapter 2

Step-By Step Application of the Johansen and Stock-Watson Approaches to Cointegration

Because the procedures developed by Johansen, Stock and Watson are more difficult to employ, this insert provides step-by-step procedures for applying these approaches. Both of these procedures can be illustrated with the multivariate model (equation (8) of the text)

$$Y_t = A_1 Y_{t-1} + A_2 Y_{t-2} + \ldots + A_p Y_{t-p} + \epsilon_t.$$

Step-by-Step Application of Johansen's Approach to Integration:

1. Pick an autoregressive order p for the model.
2. Run a regression of ΔY_t on $\Delta Y_{t-1}, \Delta Y_{t-2}, \ldots, \Delta Y_{t-p+1}$ and output the residuals, D_t. For each t, D_t has n elements.
3. Regress Y_{t-p} on $\Delta Y_{t-1}, \Delta Y_{t-2}, \ldots, \Delta Y_{t-p+1}$ and output the residuals, L_t. For each t, L_t has n elements.
4. Compute squares of the canonical correlations between D_t and L_t, calling these $\varrho_1^2 > \varrho_2^2 > \ldots > \varrho_n^2$.

5a. Letting N denote the number of time periods available in the data, compute the trace test as

$$\text{TRACE TEST } = -n \sum_{i=k+1}^{n} \ln(1 - \varrho_i^2)^1.$$

The null hypothesis is "there are k or less cointegrating vectors."

---

1. Theoretically, a cointegrating vector is associated with $\varrho_1^2, \varrho_2^2, \varrho_3^2, 0,\ 0$. The theoretical counterpart of the trace test for $H_0 : k = 3$ or less cointegrating vectors is $-N \sum_{i=4}^{s} \ln(1-0) = 0$ so the test statistic would be within sampling error of 0. For $H_0 : k = 2$ or less cointegrating vectors, the theoretical value of the test is $-N \sum_{i=3}^{s} \ln(1 - \varrho_i^2) = -N \ln(1 - \varrho_3^2) > 0$ and as N gets large, this diverges to $+\infty$. Note that ϱ_4^2 and ϱ_3^2 (which are both zero) do not contribute to the test statistic and this is the motivation for the so-called "maximal eigenvalue" test of Johansen.

5b. You may choose to use the maximal eigenvalue test (which really uses the $k+1^{\text{th}}$ maximal squared canonical correlation or eigenvalue) as follows:

$$\text{MAX EIGENVALUE TEST } = -N\ln(1-\varrho_{k+1}^2).$$

6. Compare the test statistic to the appropriate table in Johansen and Juselius (1990).[2]
   Note: The squared canonical correlations are the solution to the determinantal equation

$|\varrho_i^2 S_{kk} - S_{k0}S_{00}^{-1}\, S'_{k o}| = 0$ where $S_{kk} = N^{-1}\sum_{t=1}^{N} L_t L_t' S_{00} = N^{-1}\sum_{t=1}^{N} D_t D_t'$ and $S_{k0} = N^{-1}\sum_{t=1}^{N} L_t D_t'$ and D_t and L_t are column vectors of residuals from steps 2 and 3. The maximum likelihood estimates of the k cointegrating vectors (k columns of β_i for which $\varrho_i^2 S_{kk}\beta_i = S_{k_0}S_{00}^{-1}S'_{k0}\beta_i$).

Step-by Step Application of the Stock-Watson Approach to Cointegration:

In the Stock-Watson approach the null hypothesis is that there are m common trends ($n-k=m$) against the alternative that there are less than m, say $m-q$, common trends. There are six steps:

1. Pick the autoregressive order p for the model.
2. Compute the eigenvectors of $\Sigma Y_t Y_t'$, that is, do a principal components analysis of Y_t.
3. Using the m principal components with highest variance, that is, largest eigenvalues, fit a vector autoregression to the differences. If P_t is the vector of m principal components (select as described in the text) then the autoregressive model is denoted $\Delta P_t = A_1\Delta P_{t-1} + \ldots + A_{p-1}\Delta P_{t-p+1} + \epsilon_t$, where, as before, p stands for the number of lags in the "original" autoregressive. This provides a filter to use in step 4.
4. Compute a filtered version, $\widehat{F}_t$, of P_t by $\widehat{F}_t = P_t - \widehat{A}_1 P_{t-1} - \ldots - \widehat{A}_p P_{t-p}$. This reduces the multi-lag model to a one lag model.

2. The appropriate table as well as the handling of the intercept term in estimation depends on the role of the intercept in the model. For a discussion of this, refer to Dickey and Rossana (1990).

5. Regress $\Delta\widehat{F}_t$ on $\widehat{F}_{t-1}$ getting coefficient matrix B.
6. Compute the eigenvalues of B, normalize, and compare to the distributional tables of Stock and Watson (1988). Rejecting the null hypothesis of m common trends in favour of the alternative of $m-q$ common trends means a reduction in the number of common trends by q and thus an increase of q in the number of cointegrating vectors.

3 Unit Roots and Cointegration for the Economist

Darryl Holden and Roger Perman

3.1 INTRODUCTION

Previous papers by one of the authors, Perman (1989, 1991), have proved popular amongst applied economists seeking an introduction to the new econometrics of unit roots and cointegration. The aim of the present paper is, as before, to provide a comprehensive overview of the field in a manner which minimises the technical knowledge required of the reader and which offers intuitive explanations wherever possible. Other useful surveys, at a slightly higher technical level, include the special issues of the *Oxford Bulletin of Economics and Statistics* (March 1986, August 1992), Dolado *et al.* (1990) and Campbell and Perron (1991). In this introduction we motivate the study of unit roots and cointegration, and outline the contents of the rest of the chapter.

Economic variables, such as consumption expenditure, income and wealth, which are the variables used in our illustrative examples, are often transformed before being used in a regression analysis. A common transformation of a time series variable involves first differencing. However, the level of a variable and its first difference will typically be very different in terms of mean and variation. It is of interest, therefore, to ask whether there are formal arguments in favour of or against differencing. One approach to this issue starts by noting that estimation and hypothesis testing, using the least squares method, is justified only when the various variables being used are *stationary*. Differencing would be appealing, therefore, if the first differences of a set of variables were stationary, with the variables themselves being non-stationary. This prompts several questions: What does *stationary* mean? How can we determine whether a variable is stationary or not? How can we determine whether the first difference of a variable is stationary if the variable itself is not? These questions are dealt with in the body of the chapter.

The outline of this chapter is as follows. In section 3.2 we define and illustrate the concept of a stationary time series and contrast the statistical properties of stationary and non-stationary series. In addition we spell out the association between a series possessing a "unit root" and being "difference stationary" i.e. its first difference being a stationary series. This association allows a formal approach to the differencing issue; we need to ascertain whether the series possess a "unit root" or not. Testing for a unit root is the subject matter of section 3.3. We discuss the approaches of Dickey and Fuller (1979) and of Phillips and Perron (1988) and explain why testing for a unit root involves non-conventional statistical procedures. The tests we discuss suggest that the consumption, income and wealth series we use do possess a unit root, i.e. the variables are "difference stationary". This suggests that a regression should involve the first differences of consumption, income and wealth, rather than the levels of these variables. However, in view of the arguments of Davidson, Hendry, Srba and Yeo (1978) such a procedure can be criticised on the grounds that it ignores information about the "long run". This argument is substantiated in section 3.4 and the usefulness of the "error correction mechanism" (ECM) is explained. The proposed solution to the problem of the loss of "long run" information involves considering regressions in which both the levels of variables *and* their first differences are included. However, this leads to a tension between statistical and economic considerations. The idea of a cointegrating relationship between variables, which is the subject of section 3.5, resolves this tension. According to the ECM argument both the levels of the variables and their first differences are required so that a regression makes sense. The level terms combine to give us a "long run disequilibrium error". The circle is squared by arguing that the economic notion of a "long run disequilibrium error" and the statistical notion of a stationary time series coincide; the "long run disequilibrium error" should define a stationary series. That is, the "long run disequilibrium error" will be a combination of non-stationary variables which is itself stationary. This argument is fully developed in section 3.5, after which estimation is discussed. This involves considering the parameters in both the cointegrating relationship and the *ECM*. The different statistical properties of the variables in the cointegrating relationship and the *ECM* imply very different statistical properties for the resulting estimators. In sections 3.5.1 to 3.5.3 we discuss the estimators, obtained by least squares, which are based on the work of Engle and Granger (1987). The alternative approach of Johansen (1988) is discussed in section 3.5.4. In section 3.6 the implications of the cointegration literature for applied econometric work are explained by discussing different approaches to the determination of consumption expenditure.

The paper discusses a large number of theoretical ideas and empirical techniques which are illustrated in the Appendix to this chapter. However, two important topics are not covered in this paper. Issues involving the finite sample properties of large sample techniques are not dealt with in any detail. We have also ignored issues raised by the use of non-adjusted quarterly data. A discussion of the latter can be found in Muscattelli and Hurn (1990).

3.2 STATIONARITY AND UNIT ROOTS

3.2.1 Stationary Time Series

We begin by defining stationarity. A time series is stationary if its mean, variance and autocovariances are independent of time. Thus, suppose y_t is a time series (or stochastic process) that is defined for $t = 1, 2, \ldots$ *and* for $t = 0, -1, -2, \ldots$. Formally, y_t is covariance (weakly) stationary if the following conditions are satisfied; see Harvey (1981a, p. 22):

$$E(y_t) = \mu \tag{1}$$

$$E[(y_t - \mu)^2] = \text{var}(y_t) = \chi(0) \tag{2}$$

$$E[(y_t - \mu)(y_{t-\tau} - \mu)] = \text{cov}(y_t, y_{t-\tau}) = \chi(\tau)\ , \quad \tau = 1, 2, \ldots \tag{3}$$

Equations (1) and (2) require the process to have a constant mean and variance, while (3) requires that the covariance between any two values from the series (an autocovariance) depends only on the time interval between those two values (τ) and not on the point in time (t). The mean, variance and autocovariances are thus required to be independent of time.

For many purposes the autocorrelation function is more useful than the autocovariance function in (3) and is defined by

$$\text{corr}(y_t, y_{t-\tau}) = \frac{\text{cov}(y_t, y_{t-\tau})}{\sqrt{\text{var}(y_t)\text{var}(y_{t-\tau})}} = \frac{\chi(\tau)}{\chi(0)}, \quad \tau = 1, 2, \ldots$$

3.2.2 The First Order Autoregressive Process: AR(1)

In order to illustrate the application of the conditions for stationarity in (1), (2) and (3) we consider the process defined by

$$y_t = \rho y_{t-1} + e_t, \; t = \ldots, -1, 0, 1, \ldots \tag{4}$$

where e_t is assumed to define a sequence of independently and identically distributed (IID) random variables with expected value zero and variance σ^2. The process in 4 is stationary when ρ is less than one in absolute value, i.e. $-1 < \rho < 1$. To see this it is useful to introduce the lag operator, L, where $Ly_t = y_{t-1}$ and $L^2 y_t = L(Ly_t) = Ly_{t-1} = y_{t-2,\ldots,}$ and write the $AR(1)$ in 4 as

$$y_t - \rho y_{t-1} = y_t - \rho L y_t = (1 - \rho L) y_t = e_t \tag{5}$$

so that

$$y_t = (1 - \rho L)^{-1} e_t. \tag{6}$$

Since ρ is less than one in absolute value, we can write,

$$(1 - \rho L)^{-1} = 1 + \rho L + \rho^2 L^2 + \rho^3 L^3 + \ldots = \sum_{s=0}^{\infty} \rho^s L^s$$

which we can use in 6 to obtain

$$y_t = (1 - \rho L)^{-1} e_t = (1 + \rho L + \rho^2 L^2 + \rho^3 L^3 + \ldots) e_t$$

or

$$y_t = e_t + \rho e_{t-1} + \rho^2 e_{t-2} + \rho^3 e_{t-3} + \ldots$$

This implies that the $AR(1)$ process we are considering can be represented as a moving average process of infinite order. Given this, and the assumptions that are made about e_t in 4, it is straightforward to deduce the following results

$$E(y_t) = 0 \tag{7}$$

$$\text{var}(y_t) = \frac{\sigma^2}{1 - \rho^2} \tag{8}$$

$$\text{cov}(y_t, y_{t-\tau}) = \frac{\rho^\tau \sigma^2}{1 - \rho^2}, \quad \tau = 1, 2, \ldots \tag{9}$$

and

$$\text{corr}(y_t, y_{t-\tau}) = \rho^\tau \ , \ \tau = 1, 2, \ldots \tag{10}$$

The fact that $E(y_t)$, $\text{var}(y_t)$ and $\text{cov}(y_t, y_{t-\tau})$ do not depend on t means that the $AR(1)$ process is indeed stationary when ρ is less than one in absolute value. We refer to ρ being less than one in absolute value as the stationarity condition. This condition can be expressed in a different way if we return to equation 5 and write it in the form

$$p(L)y_t = e_t$$

where $p(L)$, $= 1 - \rho L$, is a linear function of L, the lag operator. The root of this function (i.e. the solution to $p(L) = 0$) is given by $L = (1/\rho)$, so that the requirement that ρ has absolute value less than one is equivalent to requiring that the root of $p(L)$ is greater than one in absolute value. Furthermore, $p(L)$ has a unit root, i.e. the $AR(1)$ process has a unit root, if and only if ρ is 1. In this case the stationarity condition is not satisfied. The $AR(1)$ process with a unit root is non-stationary. To explore the implications of this, we contrast the unit root ($\rho = 1$) case with the stationary case (ρ is less than one in absolute value). However, since the validity of assuming that the process starts in the infinite past is unclear when we do not assume stationarity we now assume the process starts at $t = 0$ and therefore we replace (4) with

$$y_t = \rho y_{t-1} + e_t \ , \ t = 1, 2, \ldots \tag{11}$$

where y_0 is assumed to be a fixed initial value for the process. We retain the previous assumptions as far as the e_t are concerned.

Repeated backwards substitution in 11 allows us to write

$$y_t = \rho^t y_0 + e_t + \rho e_{t-1} + \rho^2 e_{t-2} + \ldots + \rho^{t-1} e_1$$

and to obtain $E(y_t)$, $\text{var}(y_t)$ and $\text{cov}(y_t, y_{t-\tau})$ in a straightforward manner.

If we assume that $\rho = 1$, so that

$$\begin{aligned} y_t &= y_0 + e_t + e_{t-1} + e_{t-2} + \ldots + e_1 \\ &= y_{t-1} + e_t \end{aligned}$$

we obtain

$$E(y_t) = y_0$$

$$\text{var}(y_t) = t\sigma^2$$

$$\text{corr}(y_t, y_{t-\tau}) = \sqrt{\frac{t-\tau}{t}} \ , \ \tau = 1, 2, \ldots$$

whilst in the case where ρ is less than one in absolute value we have

$$E(y_t) = \rho^t y_0 \tag{12}$$

$$\mathrm{var}(y_t) = \sigma^2 \left[\frac{1-(\rho^2)^t}{1-\rho^2}\right] \tag{13}$$

$$\mathrm{corr}(y_t, y_{t-\tau}) = \rho^\tau \sqrt{\frac{1-(\rho^2)^{t-\tau}}{1-(\rho^2)^t}} \ , \ \tau = 1, 2, \ldots \tag{14}$$

There are several conclusions we can draw from these results:

(i) the case where ρ is less than one in absolute value does *not* now satisfy the stationarity conditions, since, for example, $\mathrm{var}(y_t)$ depends on t. However we still call this case the stationary case since the fact that ρ has absolute value less than one is of primary importance. Furthermore for large t the results in (12), (13) and (14) replicate those in (7), (8) and (10) in that, for (12), (13) and (14)

$$\lim_{t\to\infty} E(y_t) = 0$$

$$\lim_{t\to\infty} \mathrm{var}(y_t) = \frac{\sigma^2}{1-\rho^2}$$

$$\lim_{t\to\infty} \mathrm{corr}(y_t, y_{t-\tau}) = \rho^\tau \tag{15}$$

and in general, assuming ρ has absolute value less than one, the processes in (4) and (11) are equivalent in large samples. Thus, if $t = 1, 2, \ldots, T$ defines the values of y_t that are observed, the least squares estimator of ρ,

$$\widehat{\rho} = \frac{\sum_{t=2}^{T} y_t \ y_{t-1}}{\sum_{t=2}^{T} y_{t-1}^2} \tag{16}$$

will be such that

$$\sqrt{T}(\widehat{\rho} - \rho) \xrightarrow{d} N(0, 1-\rho^2) \tag{17}$$

($\xrightarrow{d}$ indicates convergence in distribution) whether the $AR(1)$ process starts at $t = 0$ or in the infinite past.

(ii) In the case where $\rho = 1$ the variance of y_t increases without bound as t goes to infinity and we find that

$$\lim_{t\to\infty} \mathrm{corr}(y_t, y_{t-\tau}) = 1 \ , \ \tau = 1, 2, \ldots$$

which contrasts with the result in (15) in that, as τ increases, $\mathrm{corr}(y_t, y_{t-\tau})$ fades away in the stationary case but remains constant at 1 in the unit root case. This theoretical result illustrates a common informal rule for deciding whether a times series should be considered to be stationary or not: for a stationary series the estimated autocorrelations should fade away rapidly as τ increases whereas for a non-stationary series they should not tend to.

(iii) Since

$$\frac{\partial y_t}{\partial e_{t-s}} = 1 \ , \ s = 1, 2, \ldots$$

in the unit root case, and

$$\frac{\partial y_t}{\partial e_{t-s}} = \rho^s \ , \ s = 1, 2, \ldots$$

in the stationary case, we see that a "shock", or "innovation", has a sustained effect in the unit root case and an effect that diminishes with time in the stationary case.

In the case of the $AR(1)$ process in (11) we discuss the unit root tests of Dickey and Fuller (1979, 1981) in some detail below. The tests have $\rho = 1$ as the null hypothesis and $\rho < 1$ as the alternative hypothesis and follow from Dickey and Fuller's consideration of the distribution of $\widehat{\rho}$ in (16) when $\rho = 1$. Dickey and Fuller (1979) show that the statistic

$$T(\widehat{\rho} - \rho) = T(\widehat{\rho} - 1) \tag{18}$$

has a limiting distribution. However, in contrast to (17) where the limiting distribution is normal, this limiting distribution is of a non-standard type and does *not* lead to hypothesis tests based on critical values from conventional distributions such as the normal, t and F. The critical values for unit root tests based on the limiting distribution for the statistic in (18) (or on the limiting distribution for the least squares t statistic for $\rho = 1$) have to be obtained by Dickey and Fuller using simulation techniques.

There is a further point that emerges when we contrast (17) and (18). The fact that $\widehat{\rho} - \rho$ is scaled by $\sqrt{T}$ in (17) suggests that the distribution of $\widehat{\rho} - \rho$ collapses, as T increases, at the same rate that $\sqrt{T}$ increases. The collapsing of this distribution as T increases is a reflection of the fact that $\widehat{\rho}$ is a consistent estimator of ρ. In (18) $\widehat{\rho} - 1$ is scaled by T which suggests that the distribution of $\widehat{\rho} - 1$ collapses, as T increases, at the same rate that T increases. The collapsing of this distribution as T increases is again a reflection of the fact that $\hat{\rho}$ is a consistent estimator. But the faster rate at which the second distribution collapses explains why $\hat{\rho}$ is called a "superconsistent" estimator in the case where $\rho = 1$.

The process in (11), with ($\rho = 1$), has been termed "difference stationary" since the first difference of y_t is stationary; see Nelson and Plosser (1982). This follows since, assuming $\rho = 1$,

$$y_t - y_{t-1} = \Delta y_t = e_t$$

and e_t defines a stationary process.[1]

An alternative terminology refers to a series which is itself non-stationary, but which is stationary after first differencing, as being integrated of order one, denoted $I(1)$. A series that is stationary, so that differencing is not required, is said to be integrated of order zero, denoted $I(0)$.

3.2.3 Second Order Autoregressive Case: AR(2)

Our discussion of the $AR(1)$ case can be readily extended to higher order autoregressive processes. Consider the $AR(2)$ process defined by

$$y_t = \rho_1 y_{t-1} + \rho_2 y_{t-2} + e_t$$

or

$$p(L)y_t = e_t$$

where

$$p(L) = 1 - \rho_1 L - \rho_2 L^2.$$

Since $p(L)$ is a quadratic in L it will have two roots (i.e. there are two solutions to $p(L) = 0$), r_1 and r_2 say. To discuss the stationarity conditions in this case, and to consider the implications of unit roots, it is useful to write $p(L)$ in the form

$$p(L) = (1 - \alpha_1 L)(1 - \alpha_2 L)$$

where $\alpha_1 = 1/r_1$ and $\alpha_2 = 1/r_2$, so that

$$y_t = p(L)^{-1}e_t = (1 - \alpha_1 L)^{-1}(1 - \alpha_2 L)^{-1}e_t.$$

---

1. In the notation of (1), (2) and (3) we have, for $e_t, \mu = 0, \chi(0) = \sigma^2$ and $\chi(\tau) = 0$ for $\tau = 1, 2, \ldots$

If both α_1 and α_2 are less than 1 in absolute value then we can write

$$(1-\alpha_1 L)^{-1} = 1 + \alpha_1 L + \alpha_1^2 L^2 + \alpha_1^3 L^3 + \ldots$$

and

$$(1-\alpha_2 L)^{-1} = 1 + \alpha_2 L + \alpha_2^2 L^2 + \alpha_2^3 L^3 + \ldots$$

which implies that we can write the *AR*(2) process as a moving average process of infinite order

$$\begin{aligned} y_t &= (1 + \alpha_1 L + \alpha_1^2 L^2 + \ldots)(1 + \alpha_2 L + \alpha_2^2 L^2 + \ldots)e_t \\ &= \delta_0 e_t + \delta_1 e_{t-1} + \delta_2 e_{t-2} + \ldots \end{aligned}$$

where

$$\delta_0 = 1 \ , \ \delta_1 = \alpha_1 + \alpha_2 \ , \ \delta_2 = \alpha_1^2 + \alpha_1\alpha_2 + \alpha_2^2 \ , \ \ldots$$

But it is true that any moving average process is stationary[2] so that α_1 and α_2 being less than one in absolute value are the stationarity conditions for the *AR*(2) process. Since $\alpha_1 = 1/r_1$ and $\alpha_2 = 1/r_2$ it follows that stationarity requires that r_1 and r_2, the two roots of $p(L)$, are greater than one in absolute value.[3]

Suppose now that the *AR*(2) process is non-stationary because one of the two roots of $p(L)$ is one (i.e. $p(L)$ has a single unit root). Let $r_2 = 1$ and assume r_1 is greater than one in absolute value. In this case, since $\alpha_2 = 1/r_2 = 1$, we have

$$p(L)y_t = (1-\alpha_1 L)(1-L)y_t = (1-\alpha_1 L)\Delta y_t = \Delta y_t - \alpha_1 \Delta y_{t-1} = e_t$$

---

2. As long as the required moments exist.

3. This statement remains true for the case where the roots are complex numbers. This, together with the fact that complex numbers can be represented diagrammatically as points in two dimensional space, explains the alternative statement of the stationarity conditions: that the roots of $p(L)$ are required to lie "outside the unit circle". For more on complex roots see Chiang (1984).

The *AR*(2) process we consider here can be written in an alternative form using F, the forwards shift operator, defined by $Fy_t = y_{t+1}, F^2 y_t = F(Fy_t) = F(y_{t+1}) = y_{t+2,\ldots}$

$$p^*(F)y_{t-2} = (F^2 - \rho_1 F - \rho_2)y_{t-2} = e_t$$

In fact α_1 and α_2 are the roots of $p^*(F)$ so that the stationarity conditions can be expressed in terms of $p^*(F)$ by stating that its roots are required to be less than one in absolute value (or to lie "inside the unit circle"). The generalisation of the above conditions to the case of an *AR*(p) process is discussed by Harvey (1981a, p. 32).

and the non-stationary $AR(2)$ which explains y_t implies Δy_t is generated by a stationary $AR(1)$ process. Thus y_t is $I(1)$ and Δy_t is $I(0)$ in this case. If both of the roots of $p(L)$ are one (i.e. $p(L)$ has two unit roots) then $\alpha_1 = \alpha_2 = 1$ and

$$p(L)y_t = (1-L)(1-L)y_t = (1-L)^2 y_t = \Delta^2 y_t = \Delta y_t - \Delta y_{t-1} = e_t$$

so that Δy_t is $I(1)$ and $\Delta^2 y_t$, the second difference of y_t, is $I(0)$. The y_t series needs differencing twice in order that a stationary series is obtained and is therefore $I(2)$, integrated of order two.

3.3 TESTING FOR UNIT ROOTS

3.3.1 The Dickey-Fuller Tests

As noted above Dickey and Fuller (1979, 1981) consider the $AR(1)$ process in (11) where y_0 is a fixed initial value and e_t is an IID sequence of random variables. We wish to test

$$H_0 : \rho = 1 \quad \text{against} \quad H_A : \rho < 1.$$

Under the null hypothesis y_t is non-stationary, being a random walk without drift. Under the alternative hypothesis y_t is a stationary $AR(1)$ process.

The test statistic we consider here is the standard least squares t-statistic for $\rho = 1$ although Dickey and Fuller also consider $T(\widehat{\rho} - 1)$, the standardised bias. This is straightforward to calculate and is likely to be given automatically if one estimates the reparameterisation of (11) which is obtained by subtracting y_{t-1} from both sides of the equation. This gives

$$y_t - y_{t-1} = \phi y_{t-1} + e_t \tag{19}$$

where $\phi = \rho - 1$, so that we are now interested in testing

$$H_0 : \phi = 0 \quad \textit{against} \quad H_A : \phi < 0.$$

The conventional statistic for $\phi = 0$ in 19 is identical to the t-statistic for $\rho = 1$ in (11).

Denote this t-statistic, whether obtained from (11) or 19, as t_1. We have already noted that tests of the null hypothesis of a unit root will not be based on standard distributions and that critical values have to be obtained by computer simulation. The asymptotic critical values of t_1, obtained in this manner, are given in Table 8.5.2 of Fuller (1976, p. 373).

In practice we will not be justified in using t_1 in isolation because the critical values are based on a limiting distribution which is derived under strong assumptions about e_t. It is assumed that there is no drift (i.e. no intercept term) and no time trend in the original equation. However, the limiting distribution of the test statistic and the corresponding critical values are not correct if these assumptions are false.

We first consider how to deal with the non-invariance of the Dickey Fuller test to the presence of intercepts and trends when the disturbance term is *IID*. In the next two sub-sections tests which are appropriate when the disturbance term is not *IID* will be considered.

Consider the following three models for the process determining y_t

$$y_t = \rho y_{t-1} + e_t \ , \ t = 1, 2, \ldots \tag{20.1}$$

$$y_t = \rho y_{t-1} + \alpha + e_t \ , \ t = 1, 2, \ldots \tag{20.2}$$

$$y_t = \rho y_{t-1} + \alpha + \beta t + e_t \ , \ t = 1, 2, \ldots \tag{20.3}$$

where, whichever equation is assumed to determine y_t, we assume that e_t is an *IID* process. If ρ is less than one in absolute value then (20.1) is a zero mean stationary *AR*(1) process, (20.2) is a stationary *AR*(1) process with a mean of $\alpha/(1-\rho)$, and (20.3) a stationary *AR*(1) process about a linear trend if β is non-zero.[4]

If the data are generated according to (20.1) with ρ equal to one then it can be said that y_t is integrated of order one and is a random walk without drift. If the data are generated according to (20.2) with ρ equal to one and α non-zero then y_t is again integrated of order one and is a random walk with non-zero drift. Note that in this case (20.2) can be also written in the form

$$y_t^* = y_{t-1}^* + e_t \quad \text{with} \quad y_t^* = y_t - \alpha t.$$

4. The expectations stated for (20.1) and (20.2) will apply as t goes to infinity, as discussed earlier in point (i); see section 3.2.2. Notice that, with $-1 < \rho < 1$, (20.2) and (20.3) can be written as

$$y_t^* = \rho y_{t-1}^* + e_t$$

where, for (20.3)

$$y_t^* = y_t - \frac{(\alpha - \rho(\alpha + \beta))}{(1-\rho)^2} - \frac{\beta}{1-\rho} t$$

and (setting $\beta = 0$)

$$y_t^* = y_t - \frac{\alpha}{1-\rho}$$

for (20.2).

If the data are generated according to (20.3) with ρ equal to one and β non-zero then y_t is a random walk about a non-linear time trend since (20.3) can then be written in the form

$$y_t^* = y_{t-1}^* + e_t \quad \text{with} \quad y_t^* = y_t - \left[\alpha + \frac{\beta}{2}\right] t - \frac{\beta}{2} t^2. \tag{21}$$

It is important to understand the differences between these three specifications and note the terminology used.

At this point it is useful to make explicit the various alternative combinations of estimating equation and true parameter values that can be considered. We may choose to estimate either (20.1), (20.2) or (20.3). On the other hand the values of α and β in (20.3) will necessarily be in accord with one and only one of the following possibilities

$$\alpha = 0;\ \beta = 0 \tag{I}$$
$$\alpha = 0;\ \beta \neq 0 \tag{II}$$
$$\alpha \neq 0;\ \beta = 0 \tag{III}$$
$$\alpha \neq 0;\ \beta \neq 0. \tag{IV}$$

Thus, for example, case (I) implies (20.1) is correct, (20.2) includes an unrequired intercept term and (20.3) includes both an unrequired intercept and an unrequired time trend. Case (II) implies (20.1) excludes a required time trend, (20.2) includes an unrequired intercept and excludes a required time trend and (20.3) includes an unrequired intercept.

Dickey and Fuller (1979) derive a limiting distribution for the least squares t-statistic for the null hypothesis that $\rho = 1$ where (20.1) to (20.3) are each in turn assumed to be the estimated equation, but in each case under the assumption that case (I) is correct (i.e. that (20.1) generates the data). When (20.1) is the estimated equation we denote the standard least squares t-statistic for $\rho = 1$ as t_1. When (20.2) is the estimated equation we denote the standard least squares t statistic for $\rho = 1$ as t_2, and when (20.3) is the estimated equation we denote the standard least squares t statistic for $\rho = 1$ as t_3.[5]

Each of the limiting distributions are non-standard so that computer simulation is required to obtain critical values for t_1, t_2

5. This is not the same notation as used by Dickey and Fuller who use $\widehat{\tau}$ for the statistic we call $t_1, \widehat{\tau}_\mu$ for the statistic we call t_2 and $\widehat{\tau}_\tau$ for the statistic we call t_3.

and t_3. These critical values are presented in Table 8.5.2 of Fuller (1976, p. 373). Note that because the alternative hypothesis in each case is that $\rho < 1$, a calculated value smaller than the (negative) critical value would lead to rejection of the null hypothesis of a unit root in favour of the alternative of stationarity.

The results in the last paragraph are valid in case (I). They are *not* necessarily true in the other cases we have identified. Thus:

(i) if the data are generated according to (20.2) with a non-zero intercept, i.e. case (III) describes how the data are generated, then the limiting distribution of t_2 is standard normal but the limiting distribution of t_3 remains non-standard.

(ii) if the data are generated according to (20.3) with a non-zero value of β then the limiting distributions of t_2 and t_3 are both standard normal. This is so whether or not α is zero, so that we are concerned with either case (II) or (IV).

To see the consequences of this suppose that we begin by estimating (20.3). If α and β are both zero then t_3 has the non-standard limiting distribution detailed by Dickey and Fuller. In fact the limiting distribution and the critical values remain unchanged if α is non-zero, so that we have critical values for cases (I) and (III) when (20.3) is estimated. However if β is non-zero then whatever the value of α, i.e. in cases (II) and (IV), the limiting distribution of t_3 is standard normal. The obvious problem therefore is that, unless we can decide whether or not β is zero, we will not know which critical values to use in testing for a unit root. The same type of problem occurs with the other possible estimating equations.

Since in practice we will not know the correct values for α and β, it is necessary to pursue a data-based sequential testing procedure in which one tests jointly for the presence of an intercept term and/or a time trend as well as for a unit root. This task is facilitated by the availability of a further set of critical values, tabulated in Dickey and Fuller (1981), for various statistics which are calculated as standard F test statistics and so involve comparing restricted and unrestricted residual sums of squares. We therefore have a collection of various t type and F type statistics available to us. In Table 3.1 we gather these statistics together and state the relevant null and alternative hy-

Table 3.1

A Schematic Table for Unit Root Tests

Stat.	*Eqn.*	*Null*	*Alternative*	*Distribution**	*Assumptions Required in deriving Critical Values*
t_1	20.1	$\rho = 1$	$\rho < 1$	$F(8.5.2)$ (*top*)	$\rho = 1$
t_2	20.2	$\rho = 1$	$\rho < 1$	$F(8.5.2)$ (*centre*)	$(\alpha, \rho) = (0, 1)$
t_3	20.3	$\rho = 1$	$\rho < 1$	$F(8.5.2)$ (*lower*)	$(\alpha, \beta, \rho) = (0, 0, 1)$
Φ_3	20.3	$(\alpha, \beta, \rho) = (\alpha, 0, 1)$	$(\alpha, \beta, \rho) \neq (\alpha, 0, 1)$	$DF(VI)$	$(\alpha, \beta, \rho) = (0, 0, 1)$
Φ_2	20.3	$(\alpha, \beta, \rho) = (0, 0, 1)$	$(\alpha, \beta, \rho) \neq (0, 0, 1)$	$DF(V)$	$(\alpha, \beta, \rho) = (0, 0, 1)$
Φ_1	20.2	$(\alpha, \rho) = (0, 1)$	$(\alpha, \rho) \neq (0, 1)$	$DF(IV)$	$(\alpha, \rho) = (0, 1)$

*. $F(8.5.2)$ (*top*) stands for the top panel of Table 8.5.2 in Fuller (1976). $DF(VI)$ stands for Table VI in Dickey and Fuller (1981). The other abbreviations in this column should be interpreted in a similar manner.

potheses. The way in which these statistics might be used is discussed in section 3.3.3. However before that we introduce results which extend the circumstances in which the Dickey Fuller critical values are applicable.

3.3.2 The Augmented Dickey Fuller Regression

We have so far assumed that the disturbance term, e_t, is an IID process. If this assumption is incorrect then the limiting distributions and critical values obtained by Dickey and Fuller cannot be assumed to hold. The tests of Phillips and Perron, which we shall discuss in 3.3.4, are based on a relaxation of the assumptions about the disturbance term that Dickey and Fuller work with. However Dickey and Fuller (1981) themselves demonstrate that the limiting distributions and critical values that they obtain under the assumption that e_t is an *IID* process are in fact also valid when e_t is autoregressive if the augmented Dickey Fuller (*ADF*) regression is run. Thus, assume the data are generated according to (20.1) with $\rho = 1$ and that e_t is a stationary autoregression of order p

$$e_t = \theta_1 e_{t-1} + \theta_2 e_{t-2} + \ldots + \theta_p e_{t-p} + \epsilon_t \tag{22}$$

where ϵ_t now defines an *IID* process, and consider the reparameterised version of (20.3) below,

$$\Delta y_t = \phi y_{t-1} + \alpha + \beta t + e_t \tag{23}$$

where $H_0 : \phi = 0$ is to be tested against $H_A : \phi < 0$. Given the equation for e_t in (22) we can write (23) as

$$\Delta y_t = \phi y_{t-1} + \alpha + \beta t + \theta_1 e_{t-1} + \theta_2 e_{t-2} + \ldots + \theta_p e_{t-p} + \epsilon_t$$

which can be written, since (20.1) with $\rho = 1$ gives $e_t = y_t - y_{t-1}$, as

$$\begin{aligned} \Delta y_t = \phi y_{t-1} + \alpha + \beta t + \theta_1 (y_{t-1} - y_{t-2}) \\ + \theta_2 (y_{t-2} - y_{t-3}) + \ldots + \theta_p (y_{t-p} - y_{t-p-1}) + \epsilon_t \end{aligned}$$

implying a regression of Δy_t on $y_{t-1}, \Delta y_{t-1}, \Delta y_{t-2}, \ldots, \Delta y_{t-p}$ as well as an intercept and time trend is required. Dickey and Fuller demonstrate that the t-statistic for $\phi = 0$ ($\rho = 1$ if y_t is the dependent variable) from this regression, i.e. the *ADF*(p) statistic, has the same non-standard limiting distribution if $\rho = 1$ as the statistic t_3 which was defined in the last section. This implies that the critical values for a test based on the *ADF*(p) statistic are identical to those associated with the t_3 statistic of the previous section.[6]

Said and Dickey (1984) provide a generalisation of this result. Suppose the disturbance term is an *ARMA*(p^*, q) process with p^* and q unknown. We might expect an adequate approx-

6. An alternative argument involves the reparameterisation of an *AR*(p) process for y_t. Assume, for simplicity, that $p = 2$ so that one assumes y_t is an *AR*(2)

$$p(L) y_t = e_t \quad \text{with} \quad p(L) = 1 - \rho_1 L - \rho_2 L^2$$

$p(L)$ has a unit root if and only if $\rho_1 + \rho_2 = 1$. We rewrite the *AR*(2) process

imation to this process to be obtained from an autoregression of a sufficiently high order, suggesting the *ADF* regression may remain useful. Said and Dickey confirm this in that they demonstrate that the *ADF* procedure is valid asymptotically as long as the value of p used in estimation increases at the rate $T^{1/3}$ as the sample size, T, increases.[7]

We are unlikely to know the correct value of p to use in the *ADF* regression. Therefore we have to decide this on the basis of the available data. In practice it is usual to include as many terms in the lagged dependent variable as is necessary to achieve white noise residuals. Lagrange multiplier tests can be used to test for serial correlation and thus to check whether the chosen p is adequate.

3.3.3 A Suggested Sequential Procedure for Unit Root Testing

Table 3.2 presents a decision tree summary of the test procedure discussed in this section. It is assumed an *ADF* approach is used so that sufficient lags of Δy_t are included to yield approximately white noise residuals.

The three potential estimating equations are given in (20.1) to (20.3).[8] The various tests which are used are defined in Table 3.1.

---

in the following manner

$$\begin{aligned} y_t &= \rho_1 y_{t-1} + \rho_2 y_{t-2} + e_t = (\rho_1 + \rho_2 - \rho_2) y_{t-1} + \rho_2 y_{t-2} + e_t \\ &= (\rho_1 + \rho_2) y_{t-1} - \rho_2 (y_{t-1} - y_{t-2}) + e_t \\ &= \rho y_{t-1} - \rho_2 (y_{t-1} - y_{t-2}) + e_t \end{aligned}$$

where $\rho = \rho_1 + \rho_2$ and we wish to test $\rho = 1$. The argument generalises in a straightforward manner to higher order cases. Recall that our previous discussion of the $AR(2)$ process showed that there may be two unit roots. This possibility is considered, if we accept the null hypothesis of a single unit root when the *ADF* regression is run, by repeating the process but using the first differences (rather than the level) of the series.

7. The previous paragraph assumes an $ARMA(p^*, 0)$ process with p^* a known value. We use p and p^* in order to distinguish between the true (but unknown) order of the autoregression and the order of autoregression allowed for in estimation.

8. As explained when (19) was contrasted with (11) it is often the case that the regressions are run with Δy_t as the dependent variable.

Step 1. Estimate (20.3), but include sufficient lags of Δy_t to eliminate serial correlation in the regression residuals.

Step 2. Use Φ_3 to test the null $(\alpha, \beta, \rho) = (\alpha, 0, 1)$ against the alternative $(\alpha, \beta, \rho) \neq (\alpha, 0, 1)$; see Table 3.1 for relevant sources of critical values. If the null cannot be rejected go to Step 5.

Step 3. If the null is rejected we "know" that

$$\text{either} \begin{bmatrix} \beta \neq 0 \\ \rho = 1 \end{bmatrix} \quad \text{or} \quad \begin{bmatrix} \beta = 0 \\ \rho \neq 1 \end{bmatrix} \quad \text{or} \quad \begin{bmatrix} \beta \neq 0 \\ \rho \neq 1 \end{bmatrix} \qquad (P)$$

and the next step is to test for $\rho = 1$ using the t-statistic obtained from estimating the augmented version of (20.3), with the critical values taken from the standard normal tables. Critical values from the standard normal are appropriate when β is non-zero, so that if the null hypothesis is accepted we can rule out the third case in (P). If β is zero the critical values are non-standard, but will be smaller than those obtained from the standard normal, so that an acceptance of $\rho = 1$ using standard normal critical values necessarily implies acceptance of $\rho = 1$ using non-standard critical values. Therefore the second case in (P) can also be excluded. Thus, if $\rho = 1$ is accepted when tested using standard normal critical values we conclude that β is non-zero and ρ is 1, so that the series has a unit root and a linear trend (and possibly a non-zero drift α). This outcome is highly implausible for an economic time series; see equation 21.

If we reject the null then the first alternative in (P) can be dismissed (since when β is assumed to be non-zero we reject the null that ρ is 1). This leaves the following possibilities

$$\text{either} \begin{bmatrix} \beta = 0 \\ \rho \neq 1 \end{bmatrix} \quad \text{or} \quad \begin{bmatrix} \beta \neq 0 \\ \rho \neq 1 \end{bmatrix} \qquad (P')$$

Step 4. In either case ρ is not 1, there is no unit root, and conventional test procedures can be used. Thus we may carry out a t test for the null that $\beta = 0$.

If we cannot reject this null the series is stationary with no linear trend, but possibly with an intercept. If we wish to test whether the intercept is zero, we use a conventional t test.

Table 3.2

A Step by Step Procedure for Unit Root Testing

STEP 1

Estimate the following equation

$$y_t = \alpha + \beta t + \rho y_{t-1} + \sum_{i=1}^{i=q} \theta_i \Delta y_{t-i} + e_t$$

STEP 2

Use the Φ_3 statistic to test

$$H_0 : (\alpha, \beta, \rho) = (\alpha, 0, 1) \ \nu \ H_A : (\alpha, \beta, \rho) \neq (\alpha, 0, 1)$$

If the null hypothesis is rejected go step 3.
If the null hypothesis is not rejected go to Step 5.

STEP 3

Test $\rho = 1$ using the t-statistic from step 1, with critical values from the standard normal tables.
If the null hypothesis is not rejected we conclude that β is non-zero and ρ is one.
If the null hypothesis is rejected go to Step 4.

STEP 4

Use a conventional t statistic to decide whether to take β to be zero or not.

If the null hypothesis is accepted we conclude the series is a stationary series with no trend.

If the null hypothesis is rejected we conclude the series is stationary with a linear trend.

In either case we can test hypotheses concerning the parameter α in a conventional manner.

STEP 5

Use a t statistic to test for $\rho = 1$, assuming β is zero so that non-standard critical values are required.
Assuming this t statistic provides the verification we seek we proceed to Step 6.

table continued

STEP 6

Perform a Φ_2 test for $(\alpha, \beta, \rho) = (0, 0, 1)$.
If Φ_2 leads us to conclude that α is zero
we conclude that the series is a random
walk without drift. Otherwise the series is a
random walk with drift.

In either case we can proceed to Step 7.

STEP 7

Impose the restriction that β is zero and
attempt to reestablish our conclusions concerning
ρ and α using an augmented version
of (20.2), by using the Φ_1 statistic to test
the null hypothesis of a unit root and zero drift.

If we reject this null ($\beta = 0$), the series is stationary with a linear trend, and possibly with an intercept. Again a conventional t test can be used to establish whether or not the intercept is zero.

Step 5. Given non-rejection of the null $(\alpha, \beta, \rho) = (\alpha, 0, 1)$ we "know" the series has a unit root ($\rho = 1$) with no trend ($\beta = 0$), but with possible drift.

To support the conclusion that $\rho = 1$ we may test this, given β is assumed to be zero. The required t-statistic is the same as that used in step 3 but now we need the non-standard critical values. These are found via Table 3.1, and are invariant with respect to the value of α.

Step 6. If we wish to establish whether the series has non-zero drift further tests will be required. Note that we now "know" $(\beta, \rho) = (0, 1)$ and so we might carry out the F test Φ_2 in Dickey and Fuller (1981). This tests

$$H_0 : (\alpha, \beta, \rho) = (0, 0, 1) \quad \text{against} \quad H_A : (\alpha, \beta, \rho) \neq (0, 0, 1)$$

but, given our prior information, is effectively a test for $\alpha = 0$ in this context. If we cannot reject the null, the series is a random walk without drift. If we reject the null, the series is a random walk with drift.

Step 7. We may wish to support these findings on the basis of estimating (20.2) which is obtained from (20.3) by setting β at zero as suggested by the various previous tests. If β is actually zero then tests on α and/or ρ should have greater power once this restriction is imposed.

A sensible way of proceeding might be to use the Φ_1 test for

$$H_0 : (\alpha, \rho) = (0, 1) \quad \text{against} \quad H_a : (\alpha, \rho) \neq (0, 1).$$

The calculation of this statistic is straightforward since it is an F test but the limiting distribution is non-standard. Critical values are obtained via Table 3.1.

3.3.4 Phillips and Perron Tests

The statistics proposed by Phillips and Perron (1988), termed Z statistics below, arise from their consideration of the limiting distributions of the various Dickey Fuller statistics when the assumption that e_t is an IID process is relaxed. To get a feel for the approach assume the data are generated by (20.2) and that (20.3) is the estimated equation (as in Dickey and Fuller the results below are invariant to the value of α in (20.2)). Phillips and Perron show that, given the quite general assumptions that they make concerning e_t, the statistic for $\rho = 1$ in (20.3) t_3, has a limiting distribution that involves

$$\sigma_e^2 = \lim_{T\to\infty} \left[\frac{\sum_{t=1}^{t=T} E(e_t^2)}{T}\right] \tag{24}$$

and

$$\sigma^2 = \lim_{T\to\infty} \frac{E\left[\sum_{t=1}^{t=T} e_t\right]^2}{T}. \tag{25}$$

If e_t is an IID process then σ^2 and σ_e^2 will be the same and the results of Phillips and Perron will give those previously obtained by Dickey and Fuller. However, in general, σ^2 and σ_e^2 will not be the same and the results of Dickey and Fuller, and the implied critical values for a unit root test based on t_3 will not be valid.

Phillips and Perron proposed transformations of the various statistics considered by Dickey and Fuller, for example a transformation of t_3 which we denote as $Z(t_3)$, involving the two parameters in 24 and 25 such that the Dickey Fuller limit distributions, and critical values, are restored. This means we can use the critical values given by Dickey and Fuller for the t_3 statistic if e_t is *IID*. We should, however use $Z(t_3)$ when e_t is not *IID*. To use $Z(t_3)$, and the other Z statistics of Phillips and Perron, consistent estimates of the parameters in 24 and 25 are required. Phillips and Perron propose a number of ways in which such estimates can be obtained.

Both the Phillips and Perron and the *ADF* approaches are based on asymptotic theory. Thus in both cases it is important to consider how well the limiting distributions approximate the finite sample distribution of the relevant statistic. In addition it is interesting to consider whether there is evidence concerning the relative power properties of the Phillips and Perron statistics as compared to the *ADF* statistics. Simulation evidence concerned with these issues is presented by Phillips and Perron and by Schwert (1989). Phillips and Perron assume that e_t in (20.2) is generated by a first order moving average process in their simulation experiments,

$$e_t = \epsilon_t + \theta\epsilon_{t-1}$$

where the ϵ_t are *IID*. They find that when the disturbance term has a positive moving average component (θ is positive) the power of *ADF* tests is low compared to the Phillips and Perron statistics, so that the latter statistics are preferred. However when θ is negative, matters are less clear since the evidence suggests that the Z statistics can have serious finite sample size distortions in this case.

An indication as to whether the Z statistics should be used in addition to (or in place of) the *ADF* tests might be obtained in the diagnostic statistics from the *DF* and *ADF* regressions. If normality, autocorrelation or heterogeneity statistics are significant, one might adopt the Phillips and Perron approach. Furthermore, power may be adversely affected by misspecifying the lag length in the augmented Dickey Fuller regression, although it is unclear how far this problem is mitigated by choosing the number of lags using data-based criteria, and the Z tests have the advantage that this choice does not have to be made.[9] Against this, one should avoid the use of the Z tests if the presence of negative moving average components is somehow suspected in

9. However there are alternative estimates of the quantities in 24 and 25 that one has to decide between.

the disturbances. The statistics are defined, explained and applied in Phillips and Perron (1988), and Perron (1988) and a *TSP* programme to implement the tests can be obtained from the authors. Perron (1988) concludes that the results of Nelson and Plosser, where the unit root hypothesis is accepted using augmented Dickey Fuller tests for the majority of the series under consideration, remain valid when the Phillips and Perron statistics are used.

3.3.5 Unit Root Tests and Structural Breaks

Perron (1989) considers the implications of structural breaks for the unit root tests described above. To illustrate his results consider the model defined by

$$\begin{aligned} y_t = &\alpha_1 + \beta_1 t + (\alpha_2 - \alpha_1) DU_t \\ &+ (\beta_2 - \beta_1) DT_t + e_t \ , \ t = 1, 2, \ldots, T \end{aligned} \tag{26}$$

where

$$DT_t = t \text{ if } t > TB \text{ and } 0 \text{ otherwise}$$

$$DU_t = 1 \text{ if } t > TB \text{ and } 0 \text{ otherwise}$$

and where TB is the time of the break. Both the intercept and the trend coefficient are thus allowed to change after TB.[10]

If (26) is assumed to describe how the data are generated and one looks at the distribution of the least squares estimator of ρ in the misspecified equation

$$y_t = \alpha + \beta t + \rho y_{t-1} + e_t.$$

Perron is able to show that the normalised bias $T(\widehat{\rho} - 1)$ has a probability limit that varies with $\lambda = (TB/T)$ but is always between 0 and 0.5. Since the 5% critical value for this statistic is -21.8 it follows that "the unit root hypothesis could not be rejected even asymptotically"; see Fuller (1976, p.371) for the critical values. Given this result, procedures that allow the unit root hypothesis to be tested, in the presence of structural breaks, are clearly desirable. The specification of the process to which (26) is the alternative is given by Perron as

$$y_t = \alpha_1 + \rho y_{t-1} + dD(TB)_t + (\alpha_2 - \alpha_1) DU_t + e_t \tag{27}$$

10. Other models considered are such that (i) there is simply a change in the intercept (i.e. in the level of the series) at TB and (ii) there is a change in the trend coefficient, but not in the level of the series, at TB.

with a unit root, so that $\rho = 1$. The dummy variable $D(TB)$ is defined as

$$D(TB)_t = 1 \text{ at } t = TB + 1 \text{ and } 0 \text{ otherwise}$$

so that a jump in the series at $(TB + 1)$ and a change in the drift parameter is allowed for. The equation in (27) is the unit root null hypothesis whereas the equation in (26) is the "trend stationary about a breaking trend" alternative. We wish tests of $\rho = 1$ in (27) to have high power against the alternative in (26). The estimating equation Perron proposes is obtained by pooling explanatory variables in (26) and (27) and adding lags in Δy_t, as in the augmented Dickey Fuller framework. We therefore need to estimate the parameters in the equation

$$y_t = \alpha + \theta DU_t + \beta t + \gamma DT_t + dD(TB)_t + \rho y_{t-1} + \sum_{i=1}^{i=k} c_i \Delta y_{t-i} + e_t. \quad (28)$$

Under the null hypothesis we have the restrictions $\rho = 1$ and $\gamma = \beta = 0$ whereas under the alternative hypothesis of a "trend stationary" process we expect ρ to be less than one, β, γ and θ to be non-zero and d to be close to zero. To test for the presence of a unit root using the t-statistic from the least squares estimation of (28) requires the critical values that Perron presents in Table VI.B on page 1377. If the null of a unit root is rejected then hypothesis tests concerning the remaining parameters in the estimated equation can be carried out using conventional t and F tests.

The inclusion of lags in Δy_t implies that (28) is more general than the above derivation suggests since it is, in fact, consistent with a gradual change in the intercept and trend starting at TB.

Perron illustrates the various unit root tests that allow for a structural break by reconsidering the series examined by Nelson and Plosser (1982). Using tests which did not allow for the presence of structural breaks Nelson and Plosser find that the unit root hypothesis is plausible for the majority of the series. However, as Perron notes, the series either includes 1929, the year of the Great Crash, or 1973:1, the quarter of the first oil price shock. Given this, it seems sensible to evaluate the unit root hypothesis in a setting that allows for the possibility of structural breaks. When this is done Perron finds that the hypothesis is rejected in the majority of cases (in ten out of the thirteen cases which use annual data and in the single case that involves quarterly data), and that the "trend stationary" hypothesis is more appropriate.

Perron considers the case where there is a single structural break which occurs at a known point in time. In the case where structural breaks are thought possible but one has no strong belief that a break occurs at a particular point in time the use of recursive techniques for estimation and testing seems appropriate. This issue is considered in the Appendix to this chapter.

3.3.6 Trend and Difference Stationarity

Consider the "difference stationary" equation

$$y_t = \alpha + \rho y_{t-1} + e_t \text{ with } \rho = 1 \tag{29}$$

or

$$y_t = y_0 + \alpha t + \sum_{j=1}^{t} e_j$$

as against the "trend stationary" equation

$$y_t = \mu + \beta t + e_t \tag{30}$$

where, in whichever equation we assume actually generates the data, we take e_t to satisfy the conditions of Phillips and Perron. If (30) is perceived to be a plausible alternative to (29) as an explanation of y_t then it is useful to consider the probability of rejecting the unit root hypothesis when (29) is estimated but (30) describes the data generation process. Perron (1988) has considered this issue and demonstrates that $T(\hat{\rho} - 1)$, the standardised bias, and t_2, the t statistic for $\rho = 1$, both have probability limits of zero when (30) is correct. This implies that, for a sufficiently large sample, we are essentially assured of accepting the unit root hypothesis if (30) actually describes how the data are generated, so that (29) "cannot distinguish a stationary process around a linear trend from a unit root". Perron uses this result in support of the use of (20.3) in unit root testing. The argument carries greater weight the more plausible we consider (30) to be as a potential alternative to the unit toot hypothesis, as Maddala (1992) stresses (so that a different argument would be required if we assumed the alternative to (29) was the same equation but with ρ being less than one in absolute value). In fact (30) is often considered a plausible alternative to the unit root hypothesis, and both potential explanations are obtained as special cases of (20.3). The unit root hypothesis is obtained by setting $\rho = 1$ and $\beta = 0$ whereas (30) requires an assumption that ρ has absolute value less than one since (20.3) can then be rearranged to give

$$y_t = \frac{(\alpha - \rho(\alpha + \beta))}{(1-\rho)^2} + \frac{\beta}{1-\rho}t + \epsilon_t$$

where $\epsilon_t = \rho\epsilon_{t-1} + e_t$ which will be a stationary $AR(1)$ if e_t in (20.3) is IID.

The type of argument in the previous paragraph amongst others concerning (*i*) the poor correspondence between asymptotic and finite sample distributions in simulation studies, (*ii*) the likely low power of unit root tests against the alternative of an autoregressive coefficient which is less than but close to one and (*iii*) the problems associated with the lack of invariance of the required critical values to the inclusion/exclusion of deterministic regressors (for example as considered in sections 3.3.1 and 3.3.3 and more generally in Campbell and Perron (1991)) suggests that the unit root hypothesis is not easily evaluated. In fact it has become increasingly common not to rely too heavily on pre-testing variables for their orders of integration before considering the relationship between them in a cointegration framework. However it is not clear that avoiding the prior use of unit root tests is the optimal response to the problems detailed above given that the correct orders of integration are assumed to be known in all cointegration procedures.

3.4 THE ERROR CORRECTION MECHANISM (ECM)

If a time series variable, viewed in isolation, has a single unit root then we need to take the first difference of the variable in order to obtain a stationary series. However, since economic theory is concerned with the relationship between variables, we might more usefully consider differencing within the context of a regression model. For example, let C be consumption and I be income and consider the model

$$\Delta C_t = \beta \Delta I_t + u_t \qquad (31)$$

In (31) the change in consumption from one period to the next is explained by the change in income over the same time interval.

Thus an extra pound of income per week will lead to increased expenditure of 50 pence per week if $\beta = 0.5$ without reference to any equilibrium or long run relationship between consumption and income that may exist. The decision to increase spending by 50 pence per week, given an extra pound of income, is made without considering whether current spending is too high or too low, given the current level of income. This problem with (31) is reflected in the fact that it does not have an equilibrium, or long run, solution. If we interpret equilibrium as meaning the variables become constant we want to impose

$$C_t = C_{t-1} = C_{t-2} = \dots \quad \text{and} \quad I_t = I_{t-1} = I_{t-2} = \dots$$

on (31). Doing this, and ignoring the disturbance term, gives $0 = 0$ and so does not give us an expression for C in terms of I. Such an expression, if it existed, would be the equilibrium solution of (31).

Now suppose that the value of C in equilibrium is given by C^* and assume $C^* = f(I)$. We can attempt to improve on (31) by introducing a variable that takes into account the level of C in period $(t-1)$ relative to the equilibrium value of C for the same period $(f(I_{t-1}))$. This leads to the specification

$$\Delta C_t = \beta \Delta I_t + \theta(C_{t-1} - f(I_{t-1})) + u_t$$

in which both short run and long run factors are allowed a role to play in determining how C is changed from its value in period $t-1$. The new variable is the period $t-1$ discrepancy between actual and equilibrium C and, given this, we expect θ to be a negative parameter.

If we assume that f is a linear function and write

$$C_t^* = f(I_t) = \lambda I_t$$

we obtain

$$\Delta C_t = \beta \Delta I_t + \theta(C_{t-1} - \lambda I_{t-1}) + u_t \tag{32}$$

or

$$\Delta C_t = \beta \Delta I_t + \theta C_{t-1} - \theta \lambda I_{t-1} + u_t \tag{33}$$

suggesting that a regression of ΔC_t on $\Delta I_t, C_{t-1}$ and I_{t-1} is required. A specification of this type is called an error correction mechanism and has proved popular in applied econometric research.

We have argued against the use of (31) on the grounds that

it ignores any reference to the long run aspects of decision making. On statistical grounds, however, it is straightforward. If C and I are $I(1)$, so that the first differences of these two variables are stationary, then (31), assuming it is correctly specified, will be such that conventional large sample arguments will hold. Matters are less straightforward in (32) since the variable

$$C_{t-1} - \lambda I_{t-1} \tag{34}$$

is of an uncertain status as far as stationarity is concerned. In fact since this variable is a linear combination of the variables C and I at $(t-1)$, both of which are $I(1)$, and since in general a linear combination of $I(1)$ variables is itself $I(1)$ we might conclude that the variable in (34) is $I(1)$. If this were correct it is not clear that the relationship in (32) makes sense. If the disturbance term, u_t, is assumed to be stationary ($I(0)$) then the righthand-side of the equation is a linear combination of two $I(0)$ variables and an $I(1)$ variable. Such a linear combination will be $I(1)$ but is supposedly equal to ΔC_t which is taken to be $I(0)$. Thus if $C_{t-1}\lambda I_{t-1}$ were $I(1)$ then the lack of balance between the orders of integration for the various variables in (32) suggests that that equation may be misconceived. It turns out that the idea of a cointegrating relationship between variables provides a solution to this dilemma.

3.5 COINTEGRATION[11]

Let $y_1, y_2, \ldots, y_k$ be a set of variables which we are interested in. Suppose that each variable is $I(1)$ and therefore needs differencing once to obtain a set of stationary random variables. If Y_t is the $(k,1)$ vector with i^{th} element equal to the value of y_i at time t, then Y_t is a vector of $I(1)$ variables. In general any (non-trivial) linear combination of the elements of this vector will be $I(1)$. Thus

$$\alpha' Y_t = \begin{bmatrix} \alpha_1 \\ \alpha_2 \\ \cdot \\ \cdot \\ \alpha_k \end{bmatrix}' \begin{bmatrix} y_{1t} \\ y_{2t} \\ \cdot \\ \cdot \\ y_{kt} \end{bmatrix} = \sum_{i=1}^{i=k} \alpha_i y_{it}$$

11. The definition of cointegration given here is in fact a special case of the general definition of the concept. We concentrate here on the important case as far as applied work is concerned.

(where we exclude the zero vector for α since it is trivial) is generally $I(1)$. However, there may exist linear combination(s), resulting from particular α vectors, which are $I(0)$.

We note the following points

(i) An α vector which leads to $\alpha' Y_t$ being a stationary random variable will have at least two non-zero elements. This is because we exclude the zero vector, and any α vector with a single non-zero element will result in a non-zero multiple of a single element of Y_t, which must be $I(1)$.

(ii) If we find an α vector such that $\alpha' Y_t$ is $I(0)$ then any non-trivial scalar multiple of the α vector will also lead to a stationary linear combination of the elements of Y_t.

(iii) We may be able to find a second linear combination of the elements of Y_t which is stationary and which is not related to the initial linear combination in the manner outlined above. Thus the two α vectors which we find will be linearly independent. In fact we may find up to $k-1$ linearly independent α vectors ($\alpha_1, \alpha_2, \ldots, \alpha_{k-1}$, say) such that $(\alpha_i)' Y_t$ is stationary (for $i = 1, 2, \ldots, k-1$). We can exclude the possibility of finding k linearly independent α vectors, each giving a stationary linear combination of the elements of Y_t, since this would imply that Y_t is a stationary vector.[12]

In general suppose that α_i, $i = 1, 2, \ldots, r$ $(0 < r \leq k-1)$, are linearly independent vectors such that $(\alpha_i)' Y_t$ is stationary. It follows that the (k, r) matrix

$$\alpha = [\alpha_1, \alpha_2, \ldots, \alpha_r] \tag{35}$$

will have full column rank and that the $(r, 1)$ vector

$$\alpha' Y_t = \begin{bmatrix} \alpha_1' Y_t \\ \alpha_2' Y_t \\ \cdot \\ \cdot \\ \alpha_r' Y_t \end{bmatrix}$$

12. To see this, note that the existence of k linearly independent stationary linear combinations of Y_t would imply that the matrix α, as defined in (35), is (k, k) and non-singular, with $\alpha' Y_t$ a stationary vector. But then $(\alpha')^{-1} \alpha' Y_t = Y_t$ will be a stationary vector too.

will be a vector of stationary variables. For the moment we shall leave the question of how we might determine r to one side and take it as a known constant.

We now consider an important result concerning the representation of $I(1)$ vectors which have stationary linear combinations, or which, to use the common terminology, cointegrate. The various columns of the α matrix in (35) are called cointegrating vectors. This will allow us to arrive at a natural interpretation of the cointegrating relationships. The result, the Granger Representation Theorem, establishes the existence of an *ECM* representation for such vectors, which can be written in the form

$$A(L)\Delta Y_t = -\gamma\alpha' Y_{t-1} + d(L)\epsilon_t \tag{36}$$

where ϵ_t is a $(k, 1)$ vector white noise process, such that $E(\epsilon_t) = 0$ and

$$E(\epsilon_t \epsilon_s') = \begin{matrix} G & t = s \\ 0 & t \neq s. \end{matrix}$$

In (36) $A(L)$ is a *matrix* polynomial in the lag operator and $d(L)$ is a *scalar* polynomial in the lag operator. The first matrix in $A(L)$ can be taken to be an identity matrix, so that the lefthandside in (36) can be written as

$$\Delta Y_t - A_1 \Delta Y_{t-1} - \ldots - A_p \Delta Y_{t-p}$$

assuming that $A(L)$ is of finite order (p), and

$$d(L) = 1 + d_1 L + d_2 L^2 + d_3 L^3 + \ldots$$

where $d_1, d_2, \ldots$ are scalars. This polynomial in L is applied to each element of the ϵ_t vector in (36). If $d(L)$ were 1 we would have $d(L)\epsilon_t = \epsilon_t$ in (36), a well behaved disturbance vector.

The matrix α in (36) has already been defined. The matrix γ is (k, r) and is full column rank. In fact there is an element of arbitrariness about the α and γ matrices in that we can replace α by $\alpha\Lambda$, as long as Λ is a non-singular matrix of size (r, r) and γ by $\gamma(\Lambda')^{-1}$ and leave the matrix multiplying Y_{t-1} in (36) unchanged. Replacing α by $\alpha\Lambda$ involves replacing the original r columns of α with a set of r linearly independent linear combinations of those columns. This is permissible and implies (via $\gamma(\Lambda')^{-1}$ replacing γ) a replacing of the r original columns of γ by r linearly independent linear combinations of those original columns.[13] This observation will be important when we discuss hypothesis testing.

Equation (36) and our initial discussion of $ECM's$, suggest that we should think of

$$\alpha_i' Y_t = 0 \quad i = 1, 2, \ldots, r \quad [\Longleftrightarrow \alpha' Y_t = 0]$$

as a set of r equilibrium conditions which guide the evolution of Y_t over time and

$$\mu_{it} = \alpha_i' Y_t \quad i = 1, 2, \ldots, r \tag{37}$$

as the set of r disequilibrium measures which, again as in the error correction mechanism as we initially discussed it, are required in order to obtain a convincing explanation of the change in the elements of Y_t over time. Each of these r disequilibrium measures will be involved in the explanation of each element of ΔY_t. In particular $\gamma\alpha' Y_{t-1}$ will contribute

$$\gamma_{11}(\alpha_1' Y_{t-1}) + \gamma_{12}(\alpha_2' Y_{t-1}) + \ldots + \gamma_{1r}(\alpha_r' Y_{t-1})$$

to the explanation of Δy_{1t}. In addition to this variable, (36) implies that lagged differences in all the variables $(y_1, y_2, \ldots, y_k)$ are required in explaining Δy_{1t}.

The linear combinations of Y_t which are stationary (the μ_{it} in (37)) will have zero expected values and finite variances so that not only will zero be the expected value of these terms but it will also be meaningful in that there will be a non-trivial probability of being "close" to it. In contrast if the variance were to go to infinity as the sample size increased, as occurs with a non-stationary linear combination, then the expected value would become increasingly unimportant (recall the various results in section 3.2.1). Thus, Engle and Granger (1987, p.252) argue that

> If the equilibrium concept is to have any relevance for the specification of econometric models, the economy should appear to prefer a small value for $\alpha' X_t$ rather than a large value

and translate this into a requirement that the disequilibrium measures should be zero mean and stationary.

---

13. Permissible in the sense that the matrix $\alpha\Lambda$ leads to a set of r linear combinations of Y_t that are $I(0)$ *viz.* $(\alpha\Lambda)' Y_t$.

The argument above suggests that the stationary linear combinations of the Y_t vector should be thought of as the equilibrium relationships amongst the variables because of the way in which they appear in the *ECM* representation. Further the problem we noted with respect to (32), that $C_{t-1} - \lambda I_{t-1}$ might be an $I(1)$ variable, is resolved if $r = 1$ and $C_t = \lambda I_t$ is the equilibrium or cointegrating, relationship between C and I since $C_t - \lambda I_t$ is then an $I(0)$ variable.

Whilst the *ECM* in (36) has a sound economic interpretation it appears to be rather removed from the specifications involved in unit root testing. In fact the *ECM* representation is derived from the moving average representation of ΔY_t which is similar in appearance to the unit root specifications. Since ΔY_t is a vector of stationary variables it is possible to write

$$\Delta Y_t = \pi + e_t$$

where π is a vector of drift terms and e_t can be written as a vector moving average, so that

$$e_t = C(L)\epsilon_t = \epsilon_t + C_1\epsilon_{t-1} + C_2\epsilon_{t-2} + \dots$$

where $C_i, i = 1, 2, \dots$ are (k, k) matrices and ϵ_t is a vector white noise process. It is possible, following Stock (1987), to show that Y_t is given by

$$Y_t = \pi t + C(1)\sum_{j=1}^{j=t} \epsilon_j + e_t^*$$

where

$$C(1) = I_k + C_1 + C_2 + C_3 + \dots = \sum_{j=0}^{j=\infty} C_j$$

and e_t^* is a stationary random variable, whereas

$$\sum_{j=1}^{t} \epsilon_j$$

is a vector random walk. Given this, it follows that $\alpha' Y_t$ can be written in the form

$$\alpha' Y_t = \alpha'\pi t + \alpha' C(1)\sum_{j=1}^{j=t} \epsilon_j + \alpha' e_t^*$$

and will be stationary if $\alpha'\pi$ is zero (to eliminate the trend) and $\alpha' C(1)$ is zero (to eliminate the vector random walk). This leaves the term

$$\alpha' e_t^*$$

which is a stationary random variable.

3.5.1 The Cointegrating Regression

Assume there is a single equilibrium relationship connecting the elements of Y_t. The disequilibrium measure is $\alpha' Y_t = \mu_t$, a stationary random variable. Given that we can multiply α (a vector here) by an arbitrary non-zero constant we can take any element of it and equate it to one i.e. we "normalise" α so that the chosen element is 1, and obtain something that looks like a regression model. Suppose we choose to think of the first element of α as 1 and write

$$\alpha' Y_t = y_{1t} + \alpha_2 y_{2t} + \ldots + \alpha_k y_{kt} = \mu_t.$$

This can be rearranged as

$$y_{1t} = -\alpha_2 y_{2t} - \ldots - \alpha_k y_{kt} + \mu_t.$$

This is certainly an equation we can consider estimating the parameters of by *OLS*, and is usually called the cointegrating regression. Stock (1987) has shown that the asymptotics for the *OLS* estimators in this equation are non-standard (since the variables are non-stationary). However the *OLS* estimators will be consistent. In fact they will be "super consistent", as defined in our discussion of unit root tests, since

$$T(\widehat{\beta} - \beta) \quad \text{where} \quad \beta = \begin{bmatrix} -\alpha_2 \\ -\alpha_3 \\ \cdot \\ \cdot \\ -\alpha_k \end{bmatrix}$$

will have a well defined limiting distribution. The problem is that the limiting distribution we obtain in this case is of a non-standard type and does not lead to straightforward testing procedures. This superconsistency result will apply in the example we considered earlier involving consumption and income, where the cointegrating regression is most naturally considered to involve regressing consumption on income. Furthermore, the standard argument concerning simultaneous equations bias does not apply here since it is predicated on the stationarity of the variables.

In the case where r is greater than one we should be able to choose a normalisation of the α matrix so that the first r of the k rows give us an identity matrix of size r. This leads to the estimation of r regressions in which $y_1, y_2, \ldots, y_r$ are explained in turn using $y_{r+1}, \ldots, y_k$ as explanatory variables, which, given that the labelling of the variables is arbitrary, means a set of r variables is regressed on the remaining $k - r$ variables.[14]

3.5.2 Testing for Cointegration

If we return to the case where $r = 1$ and regress y_1 on $y_2, \ldots, y_k$ we can consider the question of deciding whether or not there is a cointegrating relationship between the variables.[15] This involves examining the residuals from the cointegrating regression and in particular testing the null hypothesis that assumes the residual series has a unit root against an alternative that the series is stationary i.e. the null hypothesis is non cointegration and the alternative hypothesis is cointegration. We are thus asking the same type of question, concerning the residuals from the cointegrating regression, as we consider when we ask whether the basic variables $(y_1, y_2, \ldots, y_k)$ have a unit root, so that Dickey Fuller and augmented Dickey Fuller tests seem obvious procedures to consider. Engle and Granger consider these tests amongst others. They also consider the Durbin-Watson statistic for the cointegrating regression known as the cointegrating regression *DW* statistic or simply *CRDW*. The rationale for using this statistic lies in the work of Sargan and Bhargava (1983) which implies that the *CRDW* statistic has a probability limit of zero under the null hypothesis of non cointegration, which is therefore rejected for large calculated values of the *CRDW* statistic. Engle

14. The (k, r) matrix α has rank r. This implies that there is at least one (r, r) non-singular submatrix obtained by deleting $(k - r)$ rows. The normalisation suggested here assumes the submatrix obtained by deleting the last $(k - r)$ rows will be non-singular. The argument that any r variables can be explained using the remaining $(k - r)$ variables assumes that any (r, r) submatrix of α is non-singular.

15. A regression involving $I(1)$ variables which are not cointegrated is a "spurious" regression. The idea of a spurious regression was introduced by Granger and Newbold (1974) and subsequently developed, in the light of the cointegration literature, by Phillips (1986).

and Granger present experimental evidence in order to produce critical values for the various tests they consider and to evaluate the power of those procedures. The critical values for the Dickey Fuller statistics for a unit root in the residuals from the cointegrating regression differ from those for a unit root in the variables involved in that regression. Engle and Granger conclude that the augmented Dickey Fuller and *CRDW* tests perform best in terms of power but that the critical values for the *CRDW* are not sufficiently constant across the various experiments for the test to be practicable. Thus they conclude that the augmented Dickey Fuller test is to be preferred.

Phillips and Ouliaris (1990) present a theoretical analysis of several tests based on the residuals from the cointegrating regression. Amongst the tests considered are the augmented Dickey Fuller test, as recommended by Engle and Granger, and the transformed Dickey Fuller tests developed in the context of testing for unit roots by Phillips and Perron (the so called Z tests). Their theoretical results suggest that the Z test based on the estimated ρ from the Dickey Fuller regression is likely to have higher power than both the Z test based on the t-statistic for ρ from the Dickey Fuller regression and the augmented Dickey Fuller test. Phillips and Ouliaris give critical values for each of the test statistics, which allow for up to five variables in the cointegrating regression, and which allow for the inclusion of an intercept, and for both an intercept and time trend, in the cointegrating regression.

3.5.3 Estimating the ECM

We now consider the estimation of the short run parameters involved in the *ECM*. There are two ways we can do this. We can either use the results from the cointegrating regression and adopt a two-step approach to estimation, or we can estimate all the parameters in the *ECM* in one step, which will produce an alternative set of estimates of α to those obtained from the cointegrating regression. These procedures work as follows. Assume that $r = 1$ and that we normalise the cointegrating vector so that its first element is 1. The *ECM* in (36) implies that each element of ΔY_t is explained by the single disequilibrium variable that exists if $r = 1$, and by lagged differences in the various elements of Y_t. Given the expression for $A(L)\Delta Y_t$, as we have noted previously, we can write (36) in the form

$$\Delta y_{jt} = \sum_{i=1}^{i=k} A_{1,ji}\Delta y_{i,t-1} + \ldots + \sum_{i=1}^{i=k} A_{p,ji}\Delta y_{i,t-p} - \gamma_j(\alpha' Y_{t-1}) + d(L)\epsilon_{jt}$$

for $j = 1, 2, \ldots, k$, where $A_{1,ji}$ is, for example, the ji^{th} element of the matrix A_1. For the case $j = 1$ we obtain

$$\begin{aligned} \Delta y_{1t} &= \sum_{i=1}^{i=k} A_{1,1i}\Delta y_{i,t-1} + \ldots \\ &+ \sum_{i=1}^{i=k} A_{p,1i}\Delta y_{i,t-p} - \gamma_1(\alpha' Y_{t-1}) + d(L)\epsilon_{1t} \end{aligned} \tag{38}$$

as the equation explaining Δy_{1t}.

The two step estimation involves estimating the α vector via the cointegrating regression and then, with α set at the estimate so obtained, using *OLS* to estimate the remaining parameters in the above equation. Thus *OLS* is applied to

$$\Delta y_{1t} = \sum_{i=1}^{i=k} A_{1,1i}\Delta y_{i,t-1} + \ldots + \sum_{i=1}^{i=k} A_{p,1i}\Delta y_{i,t-p} - \gamma_1(\widehat{\alpha}' Y_{t-1}) + d(L)\epsilon_{1t}$$

where $\widehat{\alpha}'\ Y_t$ is the residual series obtained from the cointegrating regression.

The second approach to estimation treats α as an unknown vector to be estimated in (38) along with the other parameters. Given the normalisation of α that is assumed here we consider the equation

$$\begin{aligned} \Delta y_{1t} = &\sum_{i=1}^{i=k} A_{1,1i}\Delta y_{i,t-1} + \ldots + \sum_{i=1}^{i=k} A_{p,1i}\Delta y_{i,t-p} \\ &- \gamma_1(y_{1,t-1} + \alpha_2 y_{2,t-1} + \ldots + \alpha_k y_{k,t-1}) + d(L)\epsilon_{1t} \end{aligned}$$

and therefore need to regress Δy_{1t} on the various lagged differences as well as on $y_{1,t-1}, y_{2,t-1}, \ldots, y_{k,t-1}$.

Stock, together with Engle and Granger, have shown that both of the above approaches will produce consistent parameter estimates under the assumption that $d(L) = 1$. In addition Stock demonstrates that conventional asymptotic arguments are valid as far as the estimators of the short run parameters are concerned, so that the t–statistics produced by least squares can be used to test hypotheses concerning the significance of individual variables.[16]

Notice that if the α vector were known then each of the variables in (38) would be stationary, so that conventional asymptotic arguments would hold. Stock proves that the estimates of the short run parameters based on an estimated α (whether obtained from the preliminary cointegrating regression or simultaneously with the short run parameters) have the same limiting distribution as in the case where α is known. The same line of argument suggests that autocorrelation tests and the like should also be valid in this framework.

The above discussion has considered estimating the parameters in the first equation of the *ECM*. However notice that in the second approach, where α is estimated simultaneously with the short run parameters in the equation under consideration, one obtains a separate estimate of α from each of the k possible estimations. Whilst each of these estimates will be "superconsistent" it might be desirable to use a system estimator and impose the across equation parameter restrictions that are ignored if estimation proceeds on an equation by equation basis. The generalisation to the case of more than one cointegrating relationship should also be reasonably clear at this stage.

Finally notice that the above discussion treats p, the number of lags required in estimation, as a known constant. In practice we will have to determine the number of lagged differences to include on the basis of the data. This is most straightforwardly done on an equation by equation basis and we hope that by including sufficient lags we will be able to justify the assumption that the disturbance vector is well behaved.

3.5.4 Johansen Procedure

The work we have considered above leaves two basic questions unanswered: How can we test hypotheses concerning the α matrix and is it possible to test hypotheses concerning the number

16. It is important to note that this only applies to the estimates of the short run parameters. In the second approach to estimation we discussed above one also obtains estimates of the elements of the α vector. Conventional asymptotics do not hold as far as these estimates are concerned, just as in the cointegrating regression.

of equilibrium relationships? An advantage of the Johansen approach is that it enables such tests to be carried out.

The approach taken to estimation is described as Maximum Likelihood (ML) and is based on an assumption of multivariate normality, although the asymptotic arguments do not depend on this assumption. Johansen works with the ECM directly and adopts a framework that is based on the assumption that introducing sufficient lags will allow for a well behaved disturbance term. ΔY_t is assumed to be an $I(0)$ vector and the initial specification of the process involves writing

$$\Delta Y_t = \xi + \Gamma_1 \Delta Y_{t-1} + \ldots + \Gamma_{p-1} \Delta Y_{t-p+1} + \Gamma_p Y_{t-p} + \epsilon_t \qquad (39)$$

where the $(k, 1)$ vectors ϵ_t are assumed to be independently distributed as $N(0, \Lambda)$. The existence of r equilibrium relationships allow us to write

$$\Gamma_p = -\gamma\alpha'$$

and we need to consider the estimation of the following parameter matrices

$$\Gamma_1 \; \Gamma_2 \; \ldots \; \Gamma_{p-1} \; \gamma \; \alpha \; \Lambda \; \xi.$$

The vector ξ is a vector of intercepts. Its inclusion implies that there will be drift terms (π_i, $i = 1, 2, \ldots, k$), when one rearranges (39) into a form that resembles the relationships considered in unit root testing

$$y_{it} = \pi_i + y_{i,t-1} + e_{it} \qquad i = 1, 2, \ldots, k$$

or, collectively[17]

$$Y_t = \pi + Y_{t-1} + e_t.$$

Johansen (1989) shows how the π vector is related to ξ.[18]

It is of interest to consider when π is a zero vector, so that there are no drift terms allowed in the implied expression for Y_t. It turns out that $\pi = 0$ is equivalent to ξ being in the range space of the γ matrix (i.e. $\xi = \gamma v$, for some vector v) and, if we impose this in estimation, we can combine the ξ and $\Gamma_p Y_{t-p}$ terms as follows

$$\xi + \Gamma_p Y_{t-p} \;=\; \gamma\nu - \gamma\alpha' Y_{t-p} \;=\; \gamma[\nu, -\alpha'] \begin{bmatrix} 1 \\ Y_{t-p} \end{bmatrix}$$

---

17. The e_t are not IID vectors here since e_t is determined by $\epsilon_t, \epsilon_{t-1}, \epsilon_{t-2}, \ldots$

18. The analytical expression one obtains for π will satisfy the $\alpha'\pi = 0$ condition that is deduced in section 5.

which has the implication that the equilibrium relationships at time $(t-p)$ are given by

$$[\nu, -\alpha'] \begin{bmatrix} 1 \\ Y_{t-p} \end{bmatrix} = \nu - \alpha' Y_{t-p} = 0$$

and thus an intercept is included in these relationships. The restriction that $\xi = \gamma v$ $(\pi = 0)$ might be imposed in estimation if no drift terms are found in the unit root tests for the individual variables.

As far as estimation is concerned it is straightforward to obtain expressions for the ML estimators of $\Gamma_1, \Gamma_2, \ldots, \Gamma_{p-1}, \gamma, \xi$ and Λ in terms of an ML estimator of α. Johansen also shows how to obtain an estimate of α which is subject to a quite natural mathematical normalisation. The procedure is based on a given value for r, but by considering the maximised log-likelihood function (which will depend on the value of r chosen) it becomes possible to test hypotheses concerning the value of this parameter. Thus *Microfit* (1991), for example, produces a table of statistics for testing various hypotheses concerning r together with the null and alternative hypotheses and the critical values for the various tests. It should, on the basis of these results, be possible to specify a value of r.[19]

When this is done we can consider hypothesis testing. There are various types of hypothesis that can be tested involving the cointegrating vectors. The hypotheses can be tested using a likelihood ratio test (comparing restricted and unrestricted estimations) and it turns out that the limiting distribution under

19. Estimation requires that the number of lags, p, be specified. This issue is discussed in detail in the Appendix. Different values for r imply that a different subset of information is presented but there is no new estimation required. There are in fact three options in *Microfit* which produce a table of statistics for determining the value of r. One of these corresponds to the case where the restrictions required to remove drift terms are imposed in estimation. The other two cases differ only in that the critical values are different. In one case the critical values are based on the assumption that the restrictions required to remove drift terms are valid even though they are not imposed in estimation. In the second case it is assumed that these restrictions are not valid. To the extent that one does not impose the restrictions because it is felt that drift terms are required it is presumably the latter set of critical values that one requires.

the null hypothesis is chi-square, as in a conventional asymptotic analysis. It will be useful to have an example in mind. Assume that there are four variables (i.e. $k = 4$) and three equilibrium relationships (i.e. $r = 3$). We can therefore write the α matrix in the form

$$\alpha = \begin{bmatrix} \alpha_{1,1} & \alpha_{2,1} & \alpha_{3,1} \\ \alpha_{1,2} & \alpha_{2,2} & \alpha_{3,2} \\ \alpha_{1,3} & \alpha_{2,3} & \alpha_{3,3} \\ \alpha_{1,4} & \alpha_{2,4} & \alpha_{3,4} \end{bmatrix}$$

so that the three equilibrium relationships are given by

$$\begin{aligned} & \alpha_{1,1}y_{1t} + \alpha_{1,2}y_{2t} + \alpha_{1,3}y_{3t} + \alpha_{1,4}y_{4t} \\ & = \alpha_{2,1}y_{1t} + \alpha_{2,2}y_{2t} + \alpha_{2,3}y_{3t} + \alpha_{2,4}y_{4t} \\ & = \alpha_{3,1}y_{1t} + \alpha_{3,2}y_{2t} + \alpha_{3,3}y_{3t} + \alpha_{3,4}y_{4t} = 0. \end{aligned}$$

One approach to hypothesis testing involves imposing the same linear restrictions on each cointegrating vector. For example, suppose that we wish to evaluate the validity of a unit coefficient for the variable y_2 in a relationship that is normalised so that y_1 appears with a coefficient of 1, so that we wish to consider the acceptability of an equilibrium relationship of the form

$$y_{1t} = \beta y_{2t} + \gamma y_{3t} + \delta y_{4t} \tag{40}$$

with $\beta = 1$. In fact it will always be possible to select a linear combination of the three columns of the α matrix which gives *an* equilibrium relationship of this type, but what we might wish to test is whether *all* equilibrium relationships are of this type. This requires that we impose a suitable restriction on each of the three cointegrating vectors. The restriction will set the coefficient multiplying y_2 equal to the negative of the coefficient multiplying y_1. This restriction imposed on each cointegrating vector gives a total of three restrictions,

$$\alpha_{1,1} = -\alpha_{1,2} \text{ , } \alpha_{2,1} = -\alpha_{2,2} \text{ and } \alpha_{3,1} = -\alpha_{3,2}$$

and these restrictions will imply that any cointegrating vector has the property that the coefficient multiplying y_1 is the negative of the coefficient multiplying y_2, so that we get all cointegrating relationships being of the form in 40 with $\beta = 1$ when we normalise by setting the coefficient multiplying y_1 to one. The likelihood ratio test, since the same restriction is imposed on each of the three cointegrating vectors, will have three degrees of freedom in this case. In general if a set of q linear restrictions are imposed on each of the r cointegrating vectors, so as to ensure any cointegrating vector satisfies the same q linear restrictions, then the likelihood ratio test will have qr degrees of freedom. To discuss the issue of hypothesis testing further it is useful to consider a particular case. Thus, Johansen and Juselius (1990) find evidence that suggests the presence of three equilibrium relationships in a four variable data set where the variables are a money stock variable, an income variable, and two interest rates (m, y, i_1, i_2).[20]

We might speculate that these variables will be related in equilibrium by (i) a demand for money function, $m = f(y, i_1, i_2)$, (ii) an interest rate condition, $i_1 = i_2$ and (iii) an income determination equation, $y = g(m)$. Assuming relationships (i) and (iii) are linear we can write

$$m = \beta y + \gamma i_1 + \delta i_2 \qquad (i)$$

where we might wish to test whether or not β can be set at 1,

$$i_1 = i_2 \qquad (ii)$$

and

$$y = \theta m. \qquad (iii)$$

If this speculation happened to be true then we should be able to find an α matrix of the form

$$\alpha = \begin{bmatrix} m & \alpha_{1,1} & 0 & \alpha_{3,1} \\ y & \alpha_{1,2} & 0 & \alpha_{3,2} \\ i_1 & \alpha_{1,3} & \alpha_{2,3} & 0 \\ i_2 & \alpha_{1,4} & -\alpha_{2,3} & 0 \end{bmatrix}$$

where the equilibrium relationships are fed into the columns of α in an unnormalised form. We want to be able to test hypotheses concerning the estimated α matrix that allow us to assess the validity of the above speculation. Notice that if there is a unit coefficient attached to y when the money demand function (in the first column of α) is normalised so that m has a coefficient of 1, then two of three cointegrating vectors in the above matrix will satisfy the restriction we previously considered imposing on all three cointegrating vectors. However it seems unlikely that the

20. Using quarterly data for Finland.

final equilibrium relationship we identified (in the third column of α) will satisfy the same restriction. Thus imposing the same restrictions on each cointegrating vector does not seem to be useful here.

However we can also test hypotheses involving either the complete specification of a number of cointegrating vectors or the imposition of restrictions on a single cointegrating vector. In the example that we are considering our postulated equilibrium relationship between the two interest rate variables suggests we might try to fix a cointegrating vector to be of the form

$$\begin{bmatrix} 0 \\ 0 \\ 1 \\ -1 \end{bmatrix} \tag{41}$$

The requirement that an estimated cointegrating vector is of this form can be imposed in estimation and we will obtain a likelihood ratio statistic for testing the validity of imposing the restriction. The limiting null distribution will be chi-square and we therefore simply need to determine the degrees of freedom. We might suggest that this will be four since we are specifying four elements of the α matrix but this conclusion ignores the fact that the normalisation of the α matrix is arbitrary. Thus, here we can choose an appropriate linear combination of the three initial (estimated) cointegrating vectors in order to produce three new cointegrating vectors one of which will be of the form required in (41) except for one of its elements.[21]

This suggests that the likelihood ratio test for completely

21. To see this note that in the case we are considering the estimated α matrix will have rank (3). This means that at least one $(3,3)$ submatrix obtained by dropping a row will be non-singular. Assume we drop the final row and find that the matrix

$$\begin{bmatrix} \alpha_{1,1} & \alpha_{2,1} & \alpha_{3,1} \\ \alpha_{1,2} & \alpha_{2,2} & \alpha_{3,2} \\ \alpha_{1,3} & \alpha_{2,3} & \alpha_{3,3} \end{bmatrix}$$

is non-singular. However we are entitled to postmultiply any given α matrix by a $(3,3)$ non-singular matrix, Λ say, in order to obtain a new, equally valid α matrix. If we choose Λ to be such that its first column is given by

$$\begin{bmatrix} \alpha_{1,1} & \alpha_{2,1} & \alpha_{3,1} \\ \alpha_{1,2} & \alpha_{2,2} & \alpha_{3,2} \\ \alpha_{1,3} & \alpha_{2,3} & \alpha_{3,3} \end{bmatrix}^{-1} \begin{bmatrix} 0 \\ 0 \\ 1 \end{bmatrix}$$

specifying a cointegrating vector will have one degree of freedom only in this case. In general if r_1 of the r cointegrating vectors are fixed in an application involving k variables then the likelihood ratio test will have $(k-r)r_1$ degrees of freedom. In our example we have $k = 4, r = 3, r_1 = 1$ and $(k-r)r_1 = 1$.

The final thing we wish to consider is the imposition of linear restrictions on a single cointegrating vector. In our example we might wish to impose the unit income coefficient restriction on a single cointegrating vector. However our initial discussion of hypothesis testing suggested that producing a cointegrating vector of this form was a matter of normalisation rather than restriction. This observation led to our discussing the possibility of imposing this requirement on all the cointegrating vectors and implies that if we wish to impose this on a single cointegrating vector we can do so but no restriction is involved. There is nothing to test. The same conclusion would not be true where the restriction cannot be achieved by normalisation.

The above discussion should give some idea as to the types of hypothesis involving the cointegrating vectors that can be tested, and how the degrees of freedom for the tests can be determined. The importance of the fact that a normalisation, based on a view of the underlying economics, is not imposed in estimation should also be apparent.

There are various things that need to be considered before the Johansen procedure is used. The choice of p to be used in estimation is discussed in the Appendix. It is also necessary to decide how the intercept vector will be treated. That is, we need to decide whether or not we should allow for drift terms in the y_{it}, $(i = 1, 2, \ldots, k)$ variables. Recall that if we exclude the possibility of drift terms then we will find an intercept in the estimated equilibrium relationships.

It is sometimes argued that the Johansen procedure implies that the need for unit root tests is reduced. The basis for this argument is as follows: The *ECM* representation of the process determining Y_t in (39) is a reparameterisation of the vector au-

---

then the first column of the new α matrix we obtain by postmultiplication by Λ will have its first three elements exactly as we wish them to be fixed at in (41).

toregression (VAR):

$$Y_t = \xi + \Pi_1 Y_{t-1} + \ldots + \Pi_{p-1} Y_{t-p+1} + \Pi_p Y_{t-p} + \epsilon_t.$$

The stationarity conditions for vector autoregressions of this form are well known; see Harvey (1981a, p.51). These are a straightforward generalisation of the conditions in the univariate case in that they require the roots of a certain equation to be greater than one in absolute value. Unit roots in this equation imply differencing may be required in order to induce stationarity. In fact it is the case that the existence of r cointegrating relationships between the variables, as discovered via the likelihood ratio tests we discussed earlier, implies the presence of unit roots and that Y_t is $I(1)$. In this sense one might exclude unit root tests altogether if the VAR framework of Johansen is adopted.

3.6 COINTEGRATION AND ECONOMETRIC MODELLING

After the Johansen procedure is used we are likely to want to concentrate on particular elements of the ΔY_t vector, that is on particular equations in the ECM, using the estimated equilibrium relationships as explanatory variables in a conventional data based simplification ("general to specific" modelling) exercise. The idea of adopting a "general to specific," rather than "specific to general," modelling strategy, together with the idea of an error correction based approach, is perhaps most closely associated with the paper on the UK quarterly consumption function by Davidson, Hendry, Srba and Yeo (1978), hereafter $DHSY$. $DHSY$ work with non-seasonally adjusted data and argue that fourth differencing is therefore appropriate (so that, for example, decisions made in the first quarter of this year are compared to decisions made in the first quarter of last year). This leads, after the same type of arguments as in our earlier section on the error correction mechanism, to their estimating a consumption function of the form

$$\Delta_4 \ln(C_t) = \beta_1 \Delta_4 \ln(I_t) + \beta_2 \Delta_1 \Delta_4 \ln(I_t) + \beta_3 \ln \left[\frac{C}{I}\right]_{t-4} + u_t \qquad (42)$$

where C and I are consumption and income respectively. The variable

$$\ln\left[\frac{C}{I}\right]_{t-4} = \ln(C_{t-4}) - \ln(I_{t-4})$$

should be thought of as a measure of the extent to which desired and actual consumption levels differed four periods ago (i.e. in the same quarter last year). This variable together with the variables which measure the change in income are assumed to explain how consumption is changed compared to the same quarter last year.

Equation (42) is in line with the type of specifications that are of importance in the cointegration literature. Thus, if we use the notation adopted in discussing cointegration we can write

$$Y_t = \begin{bmatrix} \ln(C_t) \\ \ln(I_t) \end{bmatrix}$$

with

$$\alpha' Y_t = \alpha_1 \ln(C_t) + \alpha_2 \ln(I_t) = 0$$

in equilibrium, or, if we normalise by setting the coefficient multiplying $ln(C_t)$ at one and assume a long run unit income elasticity,

$$\ln(C_t) - \ln(I_t) = 0$$

in equilibrium. The assumption of a long run unit income elasticity by *DHSY* implies that there are no parameters to be estimated in the single equilibrium relationship connecting the variables. If this assumption were relaxed the value of this parameter might be estimated by a regression of $\ln(C_t)$ on $\ln(I_t)$, which would be the cointegrating regression.

To compare equation (42), as considered by *DHSY*, with the Johansen procedure we shall assume that $p = 4$ in the Johansen framework. If we forget the intercept vector, ξ, and normalise by setting the coefficient of $\ln(C_{t-4})$ at 1 but do not impose a long run unit income elasticity, we find that $\Delta \ln(C_t)$ can be written as a linear function of $\Delta \ln(C_{t-1})$, $\Delta \ln(C_{t-2})$, $\Delta \ln(C_{t-3})$, $\Delta \ln(I_{t-1})$, $\Delta \ln(I_{t-2})$, $\Delta \ln(I_{t-3})$ and the disequilibrium variable at $t-4$, $\ln(C_{t-4}) - \alpha_2 \ln(I_{t-4})$. Exactly the same set of explanatory variables are involved in the second equation of the system, which explains $\Delta \ln(I_t)$. But given the estimate of α_2 in the equilibrium relationship, standard asymptotics (justifying t and F tests for autocorrelation and so on) will apply if we estimate the remaining parameters in

$$\begin{aligned}\Delta\ln(C_t) = \beta_1\Delta\ln(C_{t-1}) + \beta_2\Delta\ln(C_{t-2}) + \beta_3\Delta\ln(C_{t-3})\\ \beta_4\Delta\ln(I_{t-1}) + \beta_5\Delta\ln(I_{t-2}) + \beta_6\Delta\ln(I_{t-3})\\ + \beta_7(\ln(C_{t-4}) - \widehat{\alpha}_2\ln(I_{t-4})) + \epsilon_{1t}\end{aligned} \tag{43}$$

by ordinary least squares, and one can regard this as the starting point for a standard "general to specific" specification search. Notice that this argument is very similar to that involved in the Engle–Granger two step procedure. The difference is that in the Engle–Granger case the parameters in the equilibrium relationship are estimated from the cointegrating regression whereas in the Johansen procedure the estimation technique is more complicated.

Notice that *DHSY* emphasise the use of fourth differences when non-seasonally adjusted data is used whereas the Johansen procedure is based on first differencing (the implicit assumption in the *DHSY* formulation will be that $\Delta_4\ln(C_t)$ and $\Delta_4\ln(I_t)$, as well as $\ln(C_t) - \ln(I_t)$, are stationary variables). When using *Microfit* (1991), for example, one can add seasonal dummies to the *ECM* involved in the Johansen procedure. However, in practice it seems to be the case that the Johansen procedure is used with seasonally adjusted data. Seasonal unit root tests are discussed in Charemza and Deadman (1992) and in Hylleberg (1992).

A further difference between *DHSY* and Johansen concerns the appearance of $\ln(I_t)$ in the equation explaining the change in consumption, from one quarter to the next in the case of Johansen and from one year to the next in the case of *DHSY*. This variable appears in the *DHSY* equation (42), as part of $\Delta_4\ln(I_t)$, but not in the Johansen equation 43. The inclusion of this variable raises issues concerning the exogeneity of the income variable. A full discussion of these issues is not attempted here, but we note the following points. Firstly the exogeneity issue is rather more involved than simply asking whether the correlation between $\Delta_4\ln(I_t)$ and u_t in (42) is zero or not. Following Engle, Hendry, and Richard (1983) the (weak) exogeneity of $\Delta_4\ln(I_t)$ requires that the conditional distribution of $\Delta_4\ln(C_t)$ given $\Delta_4\ln(I_t)$ provides a basis for efficient hypothesis tests concerning the parameters of interest, in that no information concerning those parameters is lost in conditioning on $\Delta_4\ln(I_t)$. The issue therefore depends on which set of parameters are of interest. If these are the parameters in the long run equilibrium relationship then a sufficient condition for the weak exogeneity of $\Delta_4 \ln(I_t)$ is that the disequilibrium error does not appear in the marginal equation that explains $\Delta_4 \ln(I_t)$. The marginal equation for $\Delta_4 \ln(I_t)$ will not involve $\Delta_4 \ln(C_t)$. (Notice that there are no long run parameters in (42) that require estimating, so that for the current discussion to be meaningful we should assume that the long run unit income elasticity restriction in (42) is relaxed). The sufficient condition is equivalent to a parameter restriction that can be tested in a number of ways. If, however, the parameters of interest are those representing the short run dynamics in (42), in addition to the long run equilibrium parameters, then sufficient conditions for the (weak) exogeneity of $\Delta_4 \ln(I_t)$ are the absence of the disequilibrium variable in the marginal equation for $\Delta_4 \ln(I_t)$, as above, and a zero correlation between the disturbance terms in the equation for $\Delta_4 \ln(C_t)$ conditional on $\Delta_4 \ln(I_t)$ and the marginal equation for $\Delta_4 \ln(I_t)$. Only this latter condition is directly related to the standard "orthogonality tests," so that these types of test are not likely to be appropriate, in isolation at least, in assessing the exogeneity of $\Delta_4 \ln(I_t)$. A fuller discussion of the above issues can be found in Urbain (1992) and Charemza and Deadman (1992).

Finally we should note that whilst (42) is an important specification in *DHSY's* paper, they actually find that it is unsatisfactory in terms of its ability to forecast accurately. They solve this problem by adding the following two variables,

$$\Delta_4 \ln(P_t) = \ln(P_t) - \ln(P_{t-4}) \quad \text{and} \quad \Delta_1\Delta_4 \ln(P_t)$$

measuring the inflation rate and the change in the inflation rate variable respectively, to obtain

$$\begin{aligned} \Delta_4 \ln(C_t) = {} & \beta_1 \Delta_4 \ln(I_t) + \beta_2 \Delta_1 \Delta_4 \ln(I_t) \\ & + \beta_3 \ln\left[\frac{C}{I}\right]_{t-4} + \beta_4 \Delta_4 \ln(P_t) + \beta_5 \Delta_1 \Delta_4 \ln(P_t) + \epsilon_t \end{aligned} \tag{44}$$

as their preferred specification. The inflation variables do *not* affect the equilibrium relationship between the variables, which is $\ln(C) = \ln(I)$ in both (42) and (44). To obtain this assume that $C_t = C_{t-1} = C_{t-2} = \ldots$, $I_t = I_{t-1} = I_{t-2} = \ldots$, and $P_t = P_{t-1} = P_{t-2} = \ldots$ so that their inclusion is presumably based on the idea that high inflation rates have only a transitional influence on consumers expenditure.

The sort of modification mentioned above is straightforward in a single equation but more work needs to be done in the case where we work with a system. Thus if we wished to make this sort of modification to an equation which included the estimated disequilibrium variable(s), as obtained from the Johansen procedure, then we would presumably have committed a misspecification because certain variables of importance (which we now wish to add to the equation under consideration) would not have been included in the original Johansen procedure. It is possible to allow for variables which play this sort of role in (39), which is the equation with respect to which we have so far discussed Johansen's work. The *ECM* with this modification will now be:

$$\Delta Y_t = \xi + \Gamma_1 \Delta Y_{t-1} + \ldots + \Gamma_{p-1} \Delta Y_{t-p+1} + \Gamma_p Y_{t-p} + \Gamma_{p+1} Z_t + \epsilon_t$$

where the Z_t vector is assumed to be made up of $I(0)$ variables, which may be differenced $I(1)$ variables. If seasonal dummies are included they will also be included as part of the Z_t vector. In the Appendix we consider an application of this framework.

3.7 CONCLUDING COMMENTS

The econometric modelling of time series using a dynamic specification which incorporates an error correction mechanism has proved fruitful and popular. Cointegration analysis may be viewed as a natural extension of this methodology and as a framework for giving such specifications a rigourous statistical foundation.

A number of alternative approaches are available to the researcher interested in estimating long run economic relationships. Two are of special interest. Vector autoregression analysis accords a very limited role to prior, theory based, restrictions, and emphasises the importance of model selection on data based criteria. Structural econometric modelling, on the other hand, emphasises the role of economic theory in the design and specification of the econometric model. Cointegration analysis can be viewed as effecting a reconciliation of these two approaches, since the existence of one or more cointegrating relationships between a set of variables implies that there are restrictions connecting the parameters in the vector autoregression. However, the dynamic relationships in both the vector autoregressive and cointegration frameworks are not subject to prior restrictions. An exposition and relative evaluation of these and other approaches to the estimation of long run economic relationships can be found in Phillips and Loretan (1991).

This paper has not considered the finite sample properties of estimators and test statistics in any systematic manner. There is an excellent survey of cointegration and unit root tests in the 1992 (August) special issue of *The Oxford Bulletin of Economics and Statistics.* A careful application of cointegration techniques to the analysis of Norwegian consumers' expenditure can be found in Brodin and Nymoen (1992). The paper is noteworthy for the particular attention given to questions concerning exogeneity and parameter stability; see, also, Drobny and Hall (1989). An application to the modelling of exchange rates is presented in MacDonald and Taylor (1992).

Appendix to Chapter 3

An Empirical Illustration

In this Appendix we illustrate the concepts and techniques introduced in the text, using the data series defined and listed in the Data Appendix at the end of this book. To avoid excessive length we have been selective in our choice of applications. As far as unit root testing is concerned, focus is placed on the use of the DF/ADF sequential search procedure. With regard to cointegration, we concentrate upon the use of the Johansen procedure for identifying the number of cointegrating relationships between a set of variables, and estimating the parameters of the long run relationships. These estimates are compared with those from the approach suggested by Engle and Granger (1987). Consideration of the short run dynamics is not undertaken here as, conditional on the presence of cointegration, the analysis of short run dynamics is based upon conventional asymptotics, and so involves little of novelty.

Denote the natural logarithms of real consumers' expenditure, real disposable income, and real personal wealth as C, I and W, noting that each time series consists of quarterly seasonally adjusted observations. Let the first differences of these series be ΔC, ΔI and ΔW respectively. Time series plots of C, I, W, ΔC, ΔI and ΔW are presented in Figures 3A.1–3A.6.

Inspection of the plots of C and I suggests linearly trended series, whereas it is much less clear whether W is trended or not. Given that the variables each appear to have a non-constant mean it appears that they are not stationary in levels. Plots of the first differences, in contrast, show no evidence of changing means nor of changing variances. This is consistent with C, I and W being integrated of order one. Such a view is supported by the correlogram evidence shown in Figures 3A.7–3A.12; whereas the estimated autocorrelations die down only slowly for the levels of C, I and W, those for the first differences damp very quickly, and then appear to fluctuate in a non-systematic way around and close to zero. We reach the tentative conclusion that C, I and W are $I(1)$ series, the first two appearing to contain drift

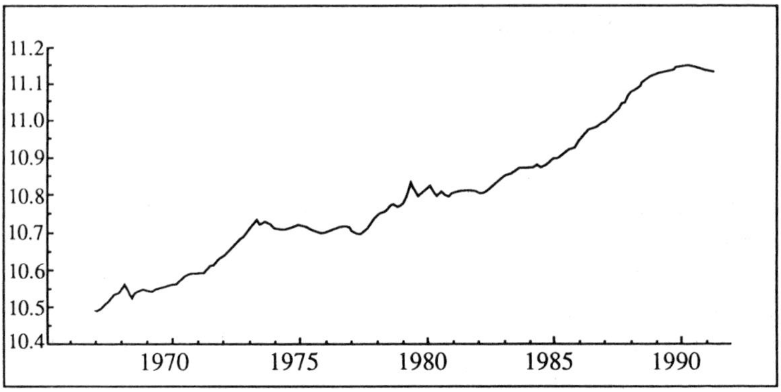

FIGURE 3A.1 Log real consumers' expenditure (C)

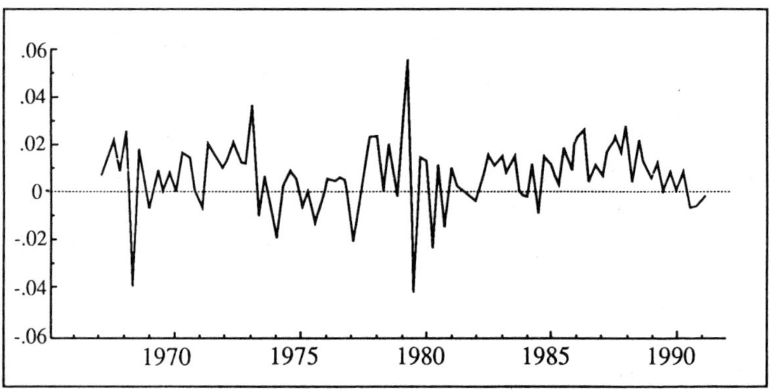

FIGURE 3A.2 Change in log real consumers' expenditure (ΔC)

components generating stochastic trends.

The use of graphical and correlogram evidence is unreliable in making inference about unit roots, and we now turn to the formal testing strategies explained in Section 3.3, examining each of the three series for the presence of unit roots.

For convenience, the parameter estimates and test statistics reported below are derived from the reparameterised versions of

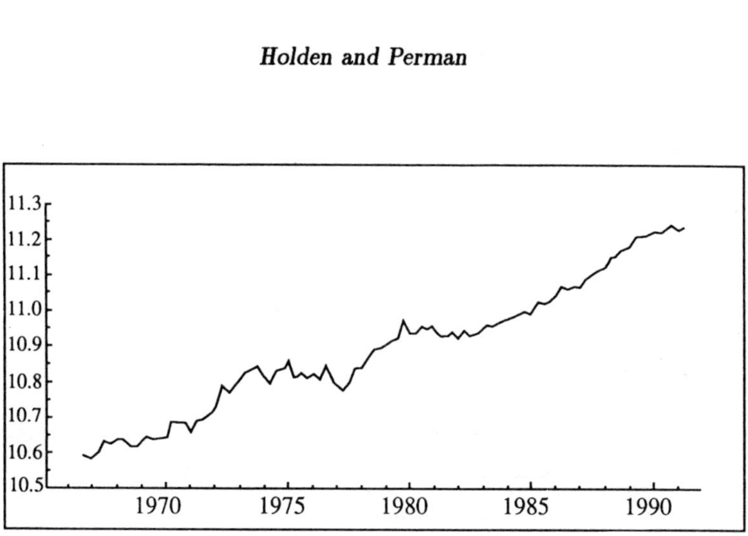

FIGURE 3A.3 Log real personal disposable income (I)

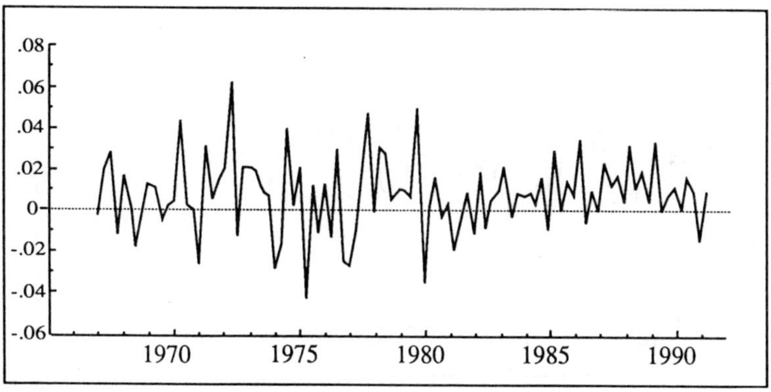

FIGURE 3A.4 Change in log real personal disposable income (ΔI)

(20.1)–(20.3) in which the dependent variable is expressed as a first difference. Unless stated otherwise, all hypothesis tests are conducted at the 5% significance level.

We begin with the consumers' expenditure time series, and estimate (20.3). To eliminate serial correlation in the residuals, three lags in the first difference of C_t were required. Line (i) in Table 3A.1 reports, therefore, the results of the $ADF(3)$ regres-

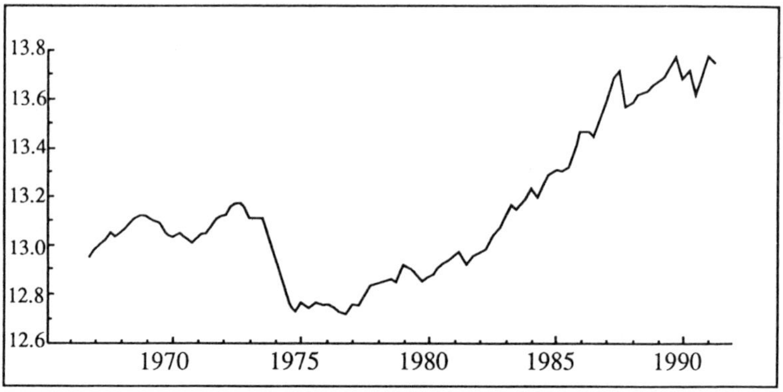

FIGURE 3A.5 Log real wealth (W)

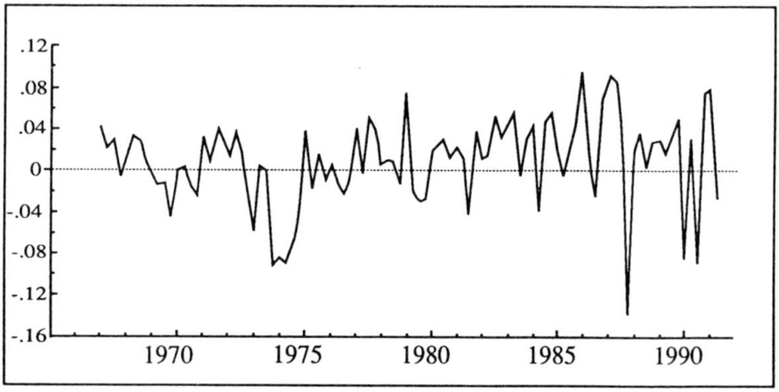

FIGURE 3A.6 Change in log real wealth (ΔW)

sion. A variable deletion test (imposing zero coefficients on C_{t-1} and the time trend) gives a computed value of 2.47 for Φ_3. Table VI in Dickey and Fuller (1981) shows the critical value for 100 observations (we actually use 96) to be 6.49. The null hypothesis cannot be rejected, implying that C contains a unit root.

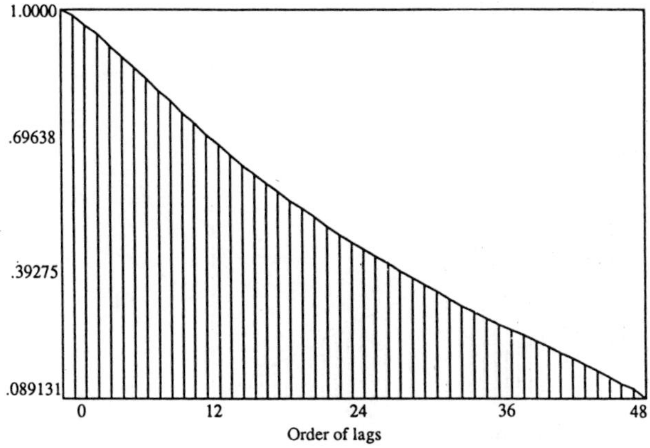

FIGURE 3A.7 Autocorrelation function of C: 1955: 1 to 1991: 2

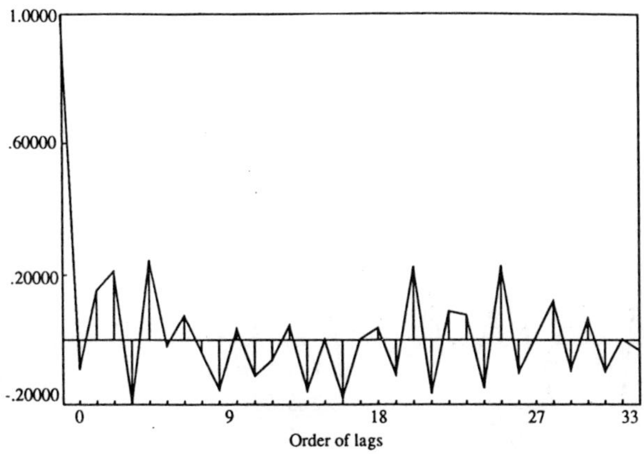

FIGURE 3A.8 Autocorrelation function of ΔC: 1966: 4 to 1991: 2

Given this decision, we move to Step 5; see Table 3.2 in the text. The calculated value of t_3, given by the t ratio of the coefficient of Z_{t-1}, is -2.22, higher than the critical value of -3.45 (obtained from Fuller, Table 8.5.2), so that our inference that the series contains a unit root is reinforced.

To ascertain whether a drift component is present, the value of the Φ_2 statistic is obtained. As the computed F statistic of

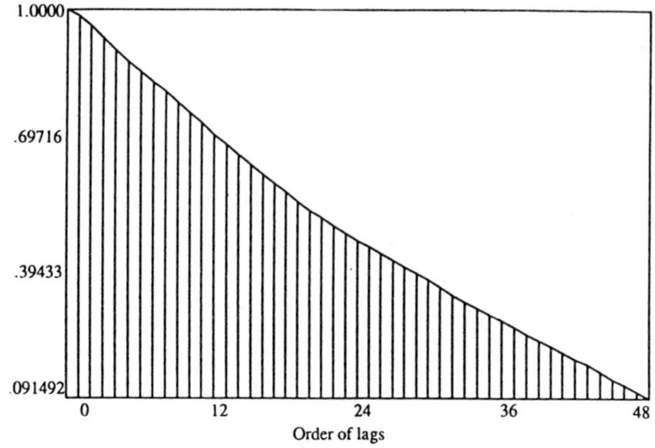

FIGURE 3A.9 Autocorrelation function of I: 1955: 1 to 1991: 2

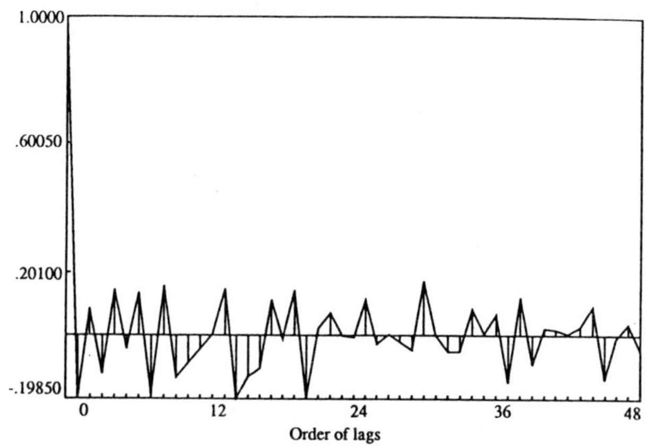

FIGURE 3A.10 Autocorrelation function of ΔI: 1955: 2 to 1991: 2

3.81 is below the tabulated critical value of 4.88, it is not possible to reject the null, which implies the absence of a drift in this process.

The information that $\beta = 0$ (from the Φ_3 test) is exploited in Step 7. Equation (20.2) is estimated (including the terms in lagged consumption) and the Φ_1 statistic is calculated; see line (ii) in Table 3A.1. The value of 3.15 is below the critical value of

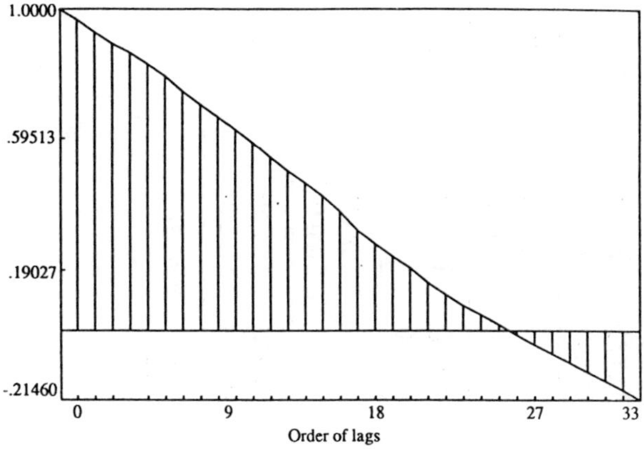

FIGURE 3A.11 Autocorrelation function of W : 1966 : 4 to 1991 : 2

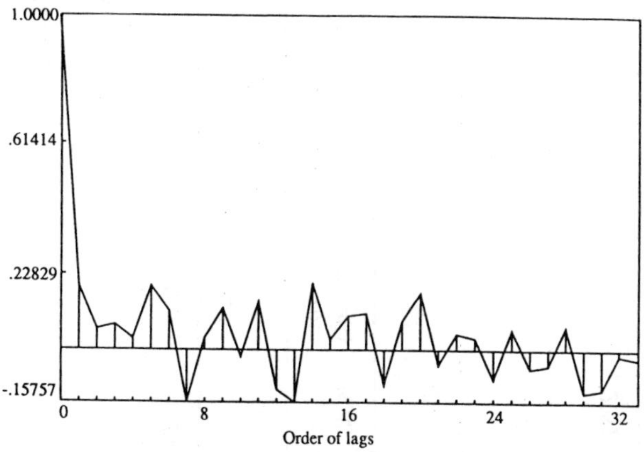

FIGURE 3A.12 Autocorrelation function of ΔW : 1967 : 1 to 1991 : 2

4.71, again leading to a decision to not reject the null.

Our conclusion from this sequence of tests is that C_t contains a unit root, but not a deterministic trend nor a drift term. However, we note that this conclusion is at odds with our previous informal analysis, which indicated the presence of non-zero drift. A more important reason for questioning the absence of drift in

TABLE 3A.1

Unit Root Tests

		Intercept	*Trend*	Z_{t-1}	ΔZ_{t-1}	ΔZ_{t-2}	ΔZ_{t-3}	$LM4$	Φ_3	Φ_2	Φ_1
		0.771	0.00048	−0.075	−0.096	0.202	0.296				
(i)	$Z = C$	(2.23)	(2.23)	(−2.22)	(−0.95)	(1.99)	(2.88)	8.47	2.47	3.81	
		0.031		−0.002	−0.132	0.164	0.256				
(ii)	$Z = C$	(0.38)		(−0.32)	(−1.30)	(1.60)	(2.48)	4.83			3.15
		1.25	0.0008	−0.121							
(iii)	$Z = I$	(2.41)	(2.41)	(−2.40)				7.44	2.93	6.27	
		0.027		−0.002							
(iv)	$Z = I$	(0.24)		(−0.18)				7.50			6.15
		0.182	0.00043	−0165							
(v)	$Z = W$	(0.76)	(2.07)	(−0.86)				2.61	2.39	2.65	
		−0.125		0.010							
(vi)	$Z = W$	(−0.67)		(0.66)				3.34			1.77

Notes:

For each regression equation the dependent variable is ΔZ_t. Z is defined in the second column of the table. The estimation period is 1967Q3 to 1991Q2 in all cases.

C_t is discussed at the end of this section.

Having ascertained that the series is not $I(0)$, one should now confirm that the series needs to be differenced only *once* to achieve stationarity (i.e. is an $I(1)$ variable). This requires further differencing (see footnote 6). Thus we begin with the regression

$$\Delta\Delta C_t = \phi\Delta C_{t-1} + \alpha + \beta t + \gamma\Delta\Delta C_{t-1} + \ldots + e_t.$$

The values of the various statistics are not reported here but we note that the results confirm the conclusion that consumption is $I(1)$.

We next examine the income process, for which test statistics are reported in Table 3A.1, rows (iii) and (iv). No lagged values of ΔI are required to achieve non serially correlated residuals in the DF regression (20.3). The Φ_3 statistic of 2.93 does not permit rejection of the null hypothesis, suggesting a unit root exists. This conclusion is reinforced by the insignificant value of t_3 which is -2.41. Calculating Φ_2 is a little less straightforward since the implied restricted model contains no variables and cannot therefore be estimated using least squares. Since, the restricted model is of the form

$$\Delta I_t = e_t$$

it is clear that the restricted residual sum of squares required in Φ_2 will be equal to $\Sigma(\Delta I_t)^2$, which equals 0.0357 in this case. Standard F test calculations then give the value of Φ_2 as 6.27. Since the relevant critical value is 4.88 the null hypothesis is rejected, implying the presence of a non-zero drift term in the income series.

The Φ_1 statistic, required in Step 7, should now lead to rejection of the null. Using the appropriate regression (line (iv) in Table 3A.1) together with the restricted residual sum of squares obtained above, we compute Φ_1 as 6.15, which exceeds the critical value of 4.71, as expected. Repeating the process with further differencing (results not reported here) confirms that only first differencing is required. Income appears to be $I(1)$.

Finally the wealth series is considered. Line (v) in Table 3A.1 reports the Dickey Fuller regression for this variable, since no lags in the dependent variable were found to be necessary. Both the t_3 and Φ_3 statistics lead to non-rejection of the relevant null, implying the presence of a unit root. The Φ_2 and Φ_1

statistics (which require manual computation, as with I) are 2.65 and 1.77 respectively, so that in neither case is the null hypothesis rejected. It appears therefore that the wealth series has a unit root, but contains neither a deterministic trend nor a drift component. Examination of the series for the presence of further unit roots suggests that single differencing only is required to achieve a stationarity series, so that W appears to be an $I(1)$ process on the basis of these tests.

Unit Root Testing in the Presence of Structural Breaks

It was argued in Section 3.3.5 that unit root tests which do not allow for the possibility of one or more structural breaks under the null and alternative hypotheses may have low power. Informal examination of the wealth series suggests the presence of level and trend shifts after 1973Q4. Some further support for the existence of a structural break after 1973Q4 is given by the results of recursive least squares estimation of the DF regression

$$\Delta W_t = \alpha + \beta t + \phi W_{t-1} + \epsilon_t$$

Figure 3A.13 shows the recursive estimates of ϕ. As the estimation period is extended through 1973Q4, the estimated coefficient on the lagged wealth term exhibits marked instability, a finding consistent with a structural break at this point.

To test for the existence of a unit root conditional on the (possible) presence of structural breaks, we use equation (28) in the text, specifying gradual shifts in level and trend to have commenced in 1973Q4. The following estimated regression is obtained, for the period 1967Q4 to 1991Q2,

$$\begin{aligned}\Delta\widehat{W}_t = \underset{(4.36)}{2.44} \underset{(-4.36)}{-0.413DU_t} \underset{(-0.66)}{-0.00062t} \underset{(3.73)}{+0.0042DT_t} \\ \underset{(0.64)}{+0.0028D(TB)_t} \underset{(-4.24)}{-0.184W_{t-1}}.\end{aligned}$$

where the least squares t statistics are given in parentheses.

The estimated regression model satisfies a full set of misspecification tests, with the exception of a test for residual normality.

Given that λ, the ratio of observations prior to the break to the full estimation sample, is (27/97), Table VI.B in Perron (1989) gives a critical value of approximately –4.21. With a calculated t statistic of –4.24 for the unit root null, we (just) reject the

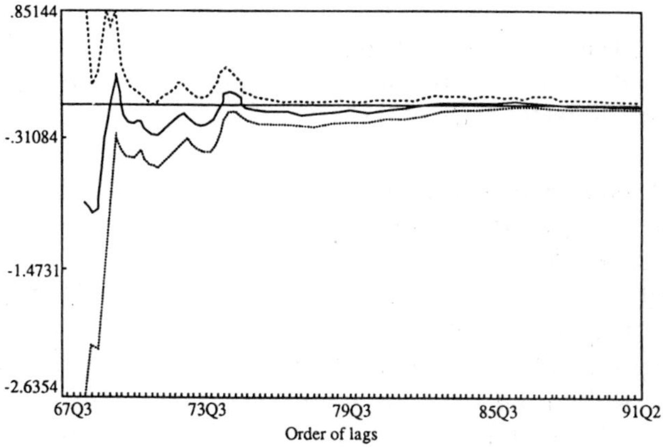

FIGURE 3A.13 Coefficient of W_{t-1} and its 2S.E. Based on Recursive OLS

null hypothesis. Furthermore, aside from the insignificant coefficient on the time trend, the parameter estimates are in accord with those expected under the alternative hypothesis. These results suggest that the wealth series does not possess a unit root, is untrended until 1973Q4 and exhibits a positive linear trend thereafter.

The above conclusion is supported by recursive estimation of the above equation, which indicates a value significantly less than one for the coefficient on W_{t-1} (and also some parameter instability indicating some remaining model misspecification).

Given that testing in a general model that permits structural breaks does not impose (what appear to be false) restrictions, one might argue that the above finding should be preferred to those from the DF tests. However, as large sample test procedures are being used we treat the evidence regarding the magnitude of the root in the wealth series as inconclusive, and thus consider the robustness of subsequent conclusions to the alternative ways in which the wealth variable might be treated.

Our conclusions to date are therefore that

(i) each series (C, I and W) contains a unit root, with only the income series having a non-zero drift term, or

(ii) C and I contain a unit root, the former without drift and the latter with drift. W is a stationary series with a deterministic trend over the latter part of its recent history.

We investigate below whether a single cointegrating relationship exists between C and I, or perhaps between C, I and W. However before we do this we need to reconsider our findings as far as the drift terms in C, I and W are concerned. Consider case (ii) (the same argument is valid for case (i)). Assume that C, I and W cointegrate and let $(\alpha_1, \alpha_2, \alpha_3)'$ be the cointegrating vector. We argued in section 3.5 that we require $\alpha'\pi$ to be zero, where π is the vector of drift terms in the moving average representation $\Delta Y_t = \pi + C(L)\epsilon_t$, where $Y' = (C, I, W)$ here. Thus, letting $\pi_i (i = 1, 2, 3)$ be the intercepts in the consumption, income and wealth equations of the system, we require

$$\alpha_1\pi_1 + \alpha_2\pi_2 + \alpha_3\pi_3 = 0.$$

Since the various DF statistics suggest $\pi_1 = \pi_3 = 0$, we require $\alpha_2\pi_2 = 0$. But the same statistics indicate that π_2 is non-zero, so that we require $\alpha_2 = 0$ which implies that income cannot enter the cointegrating relationship. This is rather unfortunate and we therefore, given our previous doubts concerning the plausibility of regarding consumption as a driftless random walk, assume that π_1 is not equal to zero.

Cointegration Analysis

Whilst it seems that C and I contain single unit roots, and so are $I(1)$ variables, the status of W is less certain. We begin by *assuming* that W is an $I(1)$ variable, in which case up to two cointegrating relationships may exist between the variables. We consider the number of cointegrating vectors that actually exist using the Johansen approach. This involves the estimation of (39) in the text with $\Gamma_p = -\gamma\alpha'$ imposed. However, we first need to select a value for p, the number of lags required in (39). This can be done in the following way.

(i) Begin with the levels equation system

$$Y_t = \xi + \Pi_1 Y_{t-1} + \ldots + \Pi_{p-1} Y_{t-p+1} + \Pi_p Y_{t-p} + \epsilon_t$$

which is a reparameterisation of the *ECM* equations in (39). We initially select a "large" value for p and estimate the above system. The system is then reestimated for successively smaller values of p. For each estimation we obtain LLF, the maximised value of the log likelihood function. (This procedure implies that Γ_p is freely estimated since the restrictions implied by the presence of cointegration are ignored).

(ii) Use likelihood ratio tests in the usual way to establish whether the restrictions imposed in moving to a reduced lag length are statistically acceptable.

(iii) Conditional upon satisfactory diagnostic test statistics for white noise residuals, select p as the shortest lag length which is consistent with the sequence of tests in (ii). Since the validity of Johansen's estimation technique is based on an assumption of white noise errors, a higher value of p than is indicated by the tests in (ii), might finally be required.

The levels systems in (i) was estimated for the sample period 1968Q4 to 1991Q2 with an initial choice of $p = 7$. An intercept term is included since non-zero drift is believed to be present in two of the series. In addition three dummy variables ($DD682$, $DD792$ and $DD883$, defined in the Data Appendix) are included in these and all subsequent regressions (although given the estimation period used in determining a value for p, $DD682$ plays no role at this point). We have no justification in economic theory for the inclusion of these dummies which were suggested by prior examination of the data. Their inclusion has two advantages. First, given their use, unambiguous results emerge from cointegration tests (which we sought for presentation simplicity in this paper). Secondly, their inclusion leads to a more satisfactory performance in terms of tests for normality in the *ECM* equation for consumer spending. The use of dummies in this way is of course open to the objection of data mining.

The following test statistics given in Table 3A., are obtained from the estimation of the levels vector autoregression in (i), for different values of p. The likelihood ratio statistics are derived in the usual way, as twice the difference between the unrestricted and restricted maximised log likelihoods. $LR(4v3)$ is, for example, the LR statistic for the nine restrictions imposed in moving from $p = 4$ to $p = 3$ (one excluded lag on each of three variables in each of three equations). With a 5% critical value of 16.92 for the $\chi^2(9)$ distribution, we are unable to reject the restrictions in moving to a lag length of $p = 2$. The LR statistic for the 45 restrictions imposed in reducing p from 7 to 2 is 54.585, which is again insignificant at the 5% level.

Table 3A.2

Test Statistics for Lag Length in the VAR

Lag length (p)	*LLF*	*LR test statistic*
1	706.325	
2	724.285	$LR(2v1) = 35.92$
3	728.786	$LR(3v2) = 9.00$
4	734.994	$LR(4v3) = 12.42$
5	741.283	$LR(5v4) = 12.56$
6	746.554	$LR(6v5) = 10.54$
7	751.577	$LR(7v6) = 10.05$

To confirm that $p = 2$ is acceptable in terms of residual whiteness, serial correlation tests are carried out using $p = 2$ in each equation of the levels vector autoregression, again using the sample period 1968Q4 to 1991Q2. Standard test statistics can be applied to the residuals from these regressions; see Lutkepohl (1991). Each statistic reported in Table 3A.3 below is the F version of the Lagrange Multiplier test for serial correlation of various orders (1, 4, 8 and 12). Since none of the statistics is significant at the 5% level, our choice of $p = 2$ is confirmed.

We now consider the use of the likelihood ratio tests of Johansen in order to determine a value for r (the number of cointegrating relationships between C, I and W) and the estimation of the matrices α and γ into which Γ_p in (39) can be factored in the presence of cointegration.

The likelihood ratio tests reported below are obtained using *Microfit* (1991). Selecting $p = 2$ and using the estimation option for trended variables, with an assumed trend in the data generating process (*DGP*) (that is, option 5 in *Microfit's* Johansen estimation menu), we obtain the statistics reported in Tables 3A.4

Table 3A.3

Test Statistics for Serial Correlation in the VAR

Dependent variable	$LM1$ $F(1,81)$	$LM4$ $F(4,78)$	$LM8$ $F(8,74)$	$LM12$ $F(12,70)$
C	0.57	2.46	1.43	1.83
I	1.00	1.54	0.94	1.00
W	0.31	0.16	1.10	1.71

and 3A.5. Table 3A.4 presents statistics which are described by *Microfit* as being based on the maximal eigenvalue. Table 3A.5 presents statistics which are described by *Microfit* as being based on the trace of stochastic matrix.

On the basis of the statistics in Tables 3A.3 and 3A.4 we conclude that $r = 1$. We wish to obtain the estimate of the single cointegrating vector that appears to exist. In normalised form this is,

$$(-1.00 \quad 0.93638 \quad 0.03804)$$

corresponding to a long run relationship of

$$C = 0.936I + 0.038W.$$

We may now test restrictions on the elements of the cointegrating vector, following the discussion of Section 3.5.4. We first test the null hypothesis that,

$$\alpha_1 = -1 \ , \ \alpha_2 = 1 \ , \ \alpha_3 = 0$$

in the equilibrium relationship,

$$\alpha_1 C + \alpha_2 I + \alpha_3 W = 0.$$

The LR statistic for this hypothesis is 4.556. The source of the critical value is the χ^2 distribution with $(k - r)r_1 = 2$ degrees of freedom; see discussion after equation (41) in the text. We do not reject the null hypothesis at the 5% significance level therefore. (In *Microfit* this statistic is obtained by selecting Option 9 in the Johansen post estimation menu, followed by Option 1, which allows us to completely specify a number of cointegrating vectors). If we test $\alpha_3 = 0$ as a single hypothesis we obtain a test statistic of 2.85, which does not lead to the null hypothesis

Johansen ML Procedure

(trended case, trend in DGP)

Cointegration LR test statistics

Sample period : 1967 : 2 *to* 1991 : 2 $p = 2$

$I(1)$ *variables included* : C, I, W.

$I(0)$ *variables included* : $DD682$, $DD792$, $DD883$.

Table 3A.4

Tests for Cointegration: 3 Variable VAR

Null	*Alternative*	*Statistic*	95%*CV*
$r = 0$	$r = 1$	26.96	20.97
$r \leq 1$	$r = 2$	5.39	14.07
$r \leq 2$	$r = 3$	2.20	3.76

Table 3A.5

Tests for Cointegration: 3 Variable VAR

Null	*Alternative*	*Statistic*	95%*CV*
$r = 0$	$r \geq 1$	34.54	29.68
$r \leq 1$	$r \geq 2$	7.58	15.41
$r \leq 2$	$r = 3$	2.20	3.76

Notes:

Table 3A.4 presents statistics which are described by *Microfit* as being based on the maximal eigenvalue. Table 3A.5 presents statistics which are described by *Microfit* as being based on the trace of stochastic matrix.

being rejected. When $\alpha_1 = -\alpha_2$ is tested as a single hypothesis we obtain a test statistic of 4.94 which does lead to the rejection of the null hypothesis (the critical value for each of the two tests comes from the $\chi^2(1)$ distribution). Notice that the restrictions $\alpha_1 = -\alpha_2$ and $\alpha_3 = 0$ are not rejected when tested jointly.

An alternative to the Johansen estimate of the cointegrating vector is obtained from the cointegrating regression, in which consumption is explained using income and wealth. The results, where an intercept is included and the estimation period is 1967Q2 to 1991Q2, are given by

$$C_t = 0.9109 I_t + 0.0790 W_t - 0.1778$$

with $R^2 = 0.99$ and $ADF(1) = -4.1358$. The ADF statistic reported here is produced automatically by *Microfit*, and is based on the ADF regression without intercept or time trend. Examination of the residuals from this regression indicates residual stationarity. The 95% critical value for $ADF(1)$ can be found in MacKinnon (1991) and is −3.8323, so that the null hypothesis (of no cointegration) is rejected. Notice that the critical values for ADF tests based on the residuals from the cointegrating regression differ from those required in undertaking unit root tests on individual variables.

The above analysis is based on an assumption that wealth is $I(1)$. As discussed earlier a plausible alternative is to assume this variable is $I(0)$. In this case we obtain the results in Tables 3A.6 and 3A.7, if we include W as an $I(0)$ variable together with the various dummies we defined earlier. The results in these tables suggest the presence of a single cointegrating relationship between C and I.

The estimated long run relationship, normalised on consumption, is given by

$$C_t = 0.94231 I_t$$

giving a point estimate of the long run income elasticity very close to those obtained previously. The null hypothesis that this elasticity is unity cannot be rejected since the required test statistic is 3.4772 and the critical value is obtained from the $\chi^2(1)$ distribution.

In the first application of the Johansen approach we assumed that wealth was an $I(1)$ variable. In the second application we assumed that wealth was an $I(0)$ variable but allowed it a role in the ECM equations for consumption and income variables (it therefore appears as an element of the Z vector in the final equation of section 3.6 of the text). A final alternative would be to exclude the wealth variable from the analysis completely. It turns out that in this case C and I still appear to cointegrate, and the Johansen procedure gives a point estimate of the equilibrium income elasticity of 0.962. Again, we are unable to reject the restriction that this elasticity is unity.

Johansen ML Procedure

(trended case, trend in DGP)

Cointegration LR test statistics

Sample period: 1967 : 2 *to* 1991 : 2 $p = 2$

$I(1)$ *variables included*: C, I.

$I(0)$ *variables included*: W $DD682$, $DD792$, $DD883$.

Table 3A.6

Tests for Cointegration: 2 Variable VAR

Null	*Alternative*	*Statistic*	95%*CV*
$r = 0$	$r = 1$	21.58	14.07
$r \leq 1$	$r = 2$	0.344	3.76

Table 3A.7

Tests for Cointegration: 2 Variable VAR

Null	*Alternative*	*Statistic*	95%*CV*
$r = 0$	$r \geq 1$	21.93	15.41
$r \leq 1$	$r = 2$	0.344	3.76

We conclude with a note of caution. A time series plot of the residuals of the cointegrating vector, adjusted for short run dynamics (see the *Microfit* manual for details) suggests some substantial departure from stationarity in the latter part of the estimation period. This is the case irrespective of which cointegrating vector we choose to believe. Whilst formal tests cannot reject cointegration, inspection of residual plots casts serious doubt on their validity.

4 Trend, Unit Root and Structural Change in Macroeconomic Time Series

Pierre Perron*

4.1 INTRODUCTION

The unit root hypothesis has attracted a considerable amount of work in both the economics and statistics literature. Indeed, the view that most economic time series are characterized by a stochastic rather than deterministic nonstationarity has become prevalent. The seminal study of Nelson and Plosser (1982) which found that most macroeconomic variables have a univariate time series structure with a unit root has catalysed a burgeoning research program with both empirical and theoretical dimensions.

As far as macroeconomic theories are concerned, the most important implication of the unit root revolution, is that under this hypothesis random shocks have a permanent effect on the system. Fluctuations are not transitory. This implication, as forcefully argued by Nelson and Plosser, has profound consequences for business cycle theories. It runs counter to the prevailing view that business cycles are transitory fluctuations around a more or less stable trend path. It is therefore important to assess carefully the reliability of the unit root hypothesis as an empirical fact.

We have recently suggested that the widespread evidence of unit roots in the univariate representation of time series may be due to the presence of important structural changes in their trend function; see Perron (1989). In that paper, we developed procedures that test for a unit root allowing the possibility of a

---

* Financial support in acknowledged from the National Science Foundation, the Fonds pour la Formation de Chercheurs de l'Aide à la Recherche du Québec and the Université de Montréal (C.A.F.I.R.).

one-time structural change in the trend function. The changes considered were of three kinds: a change in intercept (e.g. a crash), a change in slope (e.g. a productivity slow-down) or both. This simple case of a one-time change may appear overly restrictive. However, its motivation stems from its apparent relevance to many historical time series. Indeed, empirical applications using these tests on the same series analyzed by Nelson and Plosser gave evidence that weakens considerably the case for the unit root hypothesis.

The approach is in the spirit of the 'intervention analysis' of Box and Tiao (1975). According to their methodology, 'aberrant' or 'outlying' events can be separated from the noise function and be modelled as changes or 'interventions' in the deterministic part of the general time series model. Using such a strategy makes it "possible to distinguish between what can and what cannot be explained by the noise"; Box and Tiao (1975), p.72. These 'interventions' are assumed to be exogenous and to occur at a known date.

We certainly do not entertain the view that the trend function including its changes are deterministic. This would imply that one would be able to forecast with certainty future changes. This is indeed quite unappealing. What we have in mind in specifying our class of maintained hypotheses is a more general model where the intercept and the slope of the trend functions are themselves random variables modelled as integrated processes. However, the important distinction would be that the timing of the occurrence of the shocks affecting these parameters are rare relative to the sequence of innovations driving the stationary or cyclical component. The intuitive idea behind this type of modeling is that the coefficients of the trend function are determined by long-term economic fundamentals (e.g. the structure of the economic organization, population growth, etc.) and that these fundamentals are rarely changed. In this sense, the exogeneity assumption about the changes in the trend function is a device that allows to take these shocks out of the noise function into the trend function without specific modeling of the stochastic behaviour of the intercept and the slope.

This postulate concerning the exogeneity of the break date has been criticized most notably by Christiano (1988) who argued that the choice of these dates had to be viewed, to large extent, as being correlated with the data. This is an important problem because both the finite sample and asymptotic distributions of the statistics depend upon the extent of the correlation between the choice of the break points and the data. There is a sense in which the choice of these dates can be regarded as independent of the data. First, the dates reported in Perron (1989), 1929 for the Great Crash and 1973 for the start of the productivity slow-down, were chosen ex-ante and not modified ex-post. Secondly, these dates are related to exogenous events that then occurred and for which economic theory would suggest the effects that actually happened; e.g. the stock market crash of 1929 with the ensuing dismantle of the economic organization and the exogenous sudden change in oil prices with the resulting alteration of international economic coordination and policies.

In the sense described above the choice of the dates can be viewed as uncorrelated with the data. There is, however, a validity to the argument that it is only ex-post (after looking at the data) that we can say that the changes that followed these exogenous events actually occurred as predicted by the theory. Furthermore, many other exogenous events did not have the major impact that some theories would have likely predicted. In this sense, the choice of the break points must be viewed as being correlated, at least to some extent, with the data. To what extent were the choices of 1929 and 1973 correlated with the data? This is a difficult and practically impossible question to answer. At the very least, however, these choices were not perfectly correlated with the data as Perron (1989) did not systematically try various dates to maximize the chances that the unit root would be rejected; nor was there any systematic attempt to find where, according to some test criteria, are the most likely dates of change.

While the assumption about the exogeneity of the choice of the break point is probably a good first approximation to the true extent of the correlation with the data, it is useful to investigate how robust the results are to different postulates. Recent papers have addressed such issues and derived test procedures that explicitly incorporate methods to endogeneize the break date in such a way that it is fully determined by the data. This extreme case is instructive to study because if the unit root can still be rejected under such a scenario it must be the case that it would be rejected under less stringent assumptions. Papers that have addressed such issues are those of Banerjee, Lumsdaine and Stock (1992), Perron (1990b), Perron and Vogelsang (1992, 1993) and Zivot and Andrews (1992).

The aim of this paper is to review the main arguments presented in Perron (1989) and the test procedures subsequently developed with a focus on providing a comprehensive treatment useful to applied researchers. Along the way, we shall illustrate the relevance of these methods using an historical data set of real *GDP* series for eleven countries; viz. Australia, Canada, Denmark, Finland, France, Germany, Italy, Norway, Sweden, the United Kingdom and the United States. The data span the period 1870–1986 (except for Finland for which it span 1900–1986) and are the same series used by Kormendi and Meguire (1990)[1]. Empirical studies using similar series deflated by a population index (i.e. real per capita *GDP* series) include Ben-David and Papell (1992), Perron (1992) and Raj (1992). We focus here on real *GDP* series instead of their per capita counterparts since they allow us to illustrate the usefulness of a wider range of models.

The outline of the paper is as follows. Section 4.2 presents the statistical models used to analyze the unit root issue allowing for a one-time change in the slope of the trend function. Section 4.3 discusses the empirical relevance of these models in relation to the data set introduced above. Section 4.4 presents simulation results showing that standard unit root tests are biased towards non-rejection of the unit root hypothesis if the data are characterized by stationary fluctuations around a trend function that exhibits a structural change. Section 4.5 discusses the statistical procedures that allow formal tests of the unit root hypothesis.

1. The data set analyzed here was obtained through the editorial office of the *Journal of Money, Credit and Banking.* The series are annual real GDP except for the United States for which real GNP is used from the National Income and Products Accounts for the period 1929–1986, spliced to Romer's (1989) estimates for the period 1870–1928. For the United Kingdom the series are from Feinstein (1972) for the period 1870–1947 spliced to the International Financial Statistics (IFS) series of the International Monetary Fund for the period 1948–1986. For the remaining countries, the series are from Maddison (1982) spliced to the post-war IFS data. All series are analyzed using a logarithmic transformation.

We discuss, in particular, how the asymptotic critical values depend on the specific model analyzed and the method used to select the break date. Tables of critical values are provided. Section 4.6 presents empirical applications of these methods using the multi-country historical data set. Section 4.7 offers some concluding comments.

4.2 THE MODELS

The null hypothesis considered is that a given series $\{y_t\}_0^T$ (of which a sample of size $T+1$ is available) is a realization of a time series process characterized by the presence of a unit root and a possibly non-zero drift. However, the approach is generalized to allow a one-time change in the structure occurring at a time T_b $(1 < T_b < T)$. Three different models are considered under the null hypothesis: one that permits an exogenous change in the level of the series (a 'crash'), one that permits an exogenous change in the rate of growth and one that allows both changes.

As discussed in the introduction, one can view the structural changes to the trend function as some kind of "big shocks" or infrequent events that have permanent effects on the level of the series. It is important to take into consideration the way these "big shocks" affect the level of the variables, i.e. the way the transition to a new trend path occurs. In Perron (1989), two models were introduced that have different implications with respect to this transition effect. Following the terminology in Box and Tiao (1975)) the first is labelled the "additive outlier model" and specifies that the change to the new trend function occurs instantaneously. The second is labelled the "innovational outlier model" and specifies that the change to the new trend function is gradual. Of course, there is an infinity of ways that are possible, in principle, to model gradual changes following the occurrence of a "big shock". One way out of this difficulty is to suppose that the variables respond to the "big shocks" the same way as they respond to the so-called "regular shocks" (the shocks associated with the stationary noise component of the series). This is the approach taken in the modelization of the "innovational outlier model" following the treatment of intervention analyses in Box and Tiao (1975).

The distinction between the additive and innovational outlier models is important not only because the assumed transition paths are different but also because the statistical procedures to test for unit roots are also different as we shall see in Section 4.5.

4.2.1 The Additive Outlier Models

The additive outlier models are specified as follows for each of the three specifications for the types of changes occurring at the break date which we denote by T_b :

Model (1) $$y_t = \mu_1 + \beta t + (\mu_2 - \mu_1)DU_t + v_t \tag{1}$$

Model (2) $$y_t = \mu_1 + \beta_1 t + (\mu_2 - \mu_1)DU_t + (\beta_2 - \beta_1)DT_t^* + v_t \tag{2}$$

Model (3) $$y_t = \mu + \beta_1 t + (\beta_2 - \beta_1)DT_t^* + v_t \tag{3}$$

where $DU_t = 1$, $DT_t^* = t - T_b$ if $t > T_b$ and 0 otherwise. The noise component v_t is of the form $A(L)v_t = B(L)e_t$, $e_t \sim IID.$ $(0, \sigma^2)$, with $A(L)$ and $B(L)$ p^{th} and q^{th} order polynomials, respectively, in the lag operator L. The innovation series $\{v_t\}$ is taken to be of the $ARMA(p, q)$ type with the orders p and q possibly unknown. This postulate allows the series $\{y_t\}$ to represent quite general processes. The null hypothesis specifies that a root of the autoregressive polynomial is one, i.e. that we can write $A(L) = (1 - L)A*(L)$ where all the roots of $A*(L)$ are outside the unit circle. Under the alternative hypothesis of stationary fiuctuations around the trend function all the roots of $A(L)$ are strictly outside the unit circle. Note that the changes in the trend function are allowed to occur under both the null and alternative hypotheses.

Model (1) describes what we shall refer to as the crash model. It allows for a one-time change in the intercept of the trend function. Model (2) allows both a change in the intercept and the slope of the trend function to take place simultaneously, i.e. a sudden change in level followed by a different growth path. As in Model (1), the segments of the trend function are, in general, not joined at the time of break. Model (3) is referred to as the "changing growth" model. It allows for a change in the slope of the trend function without any sudden change in the level at the time of the break. This implies that the two segments of the trend function are joined at the time of break.

4.2.2 The Innovational Outlier Models

The innovational outlier models are easier to characterize by describing them separately under the null and alternative hypotheses. Note also that the innovational outlier versions have been considered only for Models (1) and (2). The basic reason is that the innovational outlier version of Model (3) does not lend itself easily to empirical applications using linear estimation methods.

We recall from the above discussion that the innovational outlier model specifies that the "shocks" (or changes) to the trend function affect the level of the series the same way as the "regular shocks" do. The temporal effect of random shocks is easily analyzed using the moving average representation of the stochastic component. Under the null hypothesis, this leads us to the following representations:

Model (1) $\quad y_t = y_{t-1} + b + \psi(L)(e_t + \delta D(T_b)_t),$

Model (2) $\quad y_t = y_{t-1} + b + \psi(L)(e_t + \delta D(T_b)_t + \eta DU_t),$

where $D(T_b)_t = 1$ if $t = T_b + 1$ and 0 otherwise. Here, the moving average representation of the first-differences of the data is specified by the lag polynomial $\psi(L)$ (possibly infinite with $\psi(0) = 1$). That is, denoting by z_t the noise function of the series, we have $A_1(L)z_t = B_1(L)e_t$ and $\psi(L) = A_1(L)^{-1}B_1(L)$. Throughout, it is assumed that the finite order polynomials $A_1(L)$ and $B_1(L)$ have all their roots outside the unit circle. Under this specification we have, for Model (1), that the immediate impact of the change in the intercept is δ while the long run impact is $\psi(1)\delta$. Similarly, under Model (2), the immediate impact of the change in slope is η while the long run impact is $\psi(1)\eta$.

Under the alternative hypothesis of stationary fluctuations, the models are:

Model (1) $\quad y_t = \mu + \beta t + \Phi(L)(e_t + \theta DU_t) \qquad (4)$

Model (2) $\quad y_t = \mu + \beta t + \Phi(L)(e_t + \theta DU_t + \gamma DT_t^*) \qquad (5)$

where $\Phi(L) = (1 - \alpha L)^{-1}A_1(L)^{-1}B_1(L)$ with $A_1(L)$ and $B_1(L)$ as defined before. In Models (1) and (2), the immediate impact of the change in the intercept of the trend function is θ while the long run impact is $\Phi(1)\theta$. Similarly, in Model (2), the immediate impact of the change in slope is γ while the long run impact is $\Phi(1)\gamma$.

The models under the null and alternative hypotheses can be nested in the following way. For Model (1), we have:

$$y_t = \mu + \theta DU_t + \beta t + \delta D(TB)_t + \alpha y_{t-1} + \sum_{i=1}^{\infty} c_i \Delta y_{t-i} + e_t \tag{6}$$

and for Model (2)

$$y_t = \mu + \theta DU_t + \beta t + \gamma DT_t^* + \delta D(TB)_t + \alpha y_{t-1} + \sum_{i=1}^{\infty} c_i \Delta y_{t-i} + e_t. \tag{7}$$

The coefficient c_i in (6) and (7) correspond to the autoregressive representation of the moving-average polynomial $A_1(L)^{-1}B_1(L)$ given by $\phi(L) = (c_1 L + c_2 L^2 + \cdots) = \psi(L)^{-1} = B_1(L)^{-1}A_1(L)$. The polynomial $\phi(L)$ will be of infinite order if moving average components are present, i.e. if $B_1(L) \neq 1$. In practice, this infinite sum will, of course, be approximated by a finite sum in order to derive operational tests.

The null hypothesis imposes the following restrictions on the coefficients. For model (6), these are $\alpha = 1, \theta = \beta = 0$ and, in general, $\delta \neq 0$ (if there is a change in the intercept). For Model (7), the restrictions are $\alpha = 1, \beta = \gamma = 0$ and again, in general, $\delta \neq 0$. Under the alternative hypothesis, we have the following specifications: $|\alpha| < 1$ and, in general, $\delta = 0$ in both Models (6) and (7). These restrictions are, however, not imposed by the testing procedures described latter.

4.3 MOTIVATION

To motivate the models presented in Section 4.2, we consider the historical data set of real *GDP* series described in the introduction. Figure 4.1 presents the graph of each series. A marked characteristic is first the noticeable pattern of increase through time. Hence, the series can be said to be trending. However, this increase over time does not, for most countries, follow a steady pattern. Instead, many series are seen to exhibit one or more of the following features: (i) a more or less sudden decrease in level (e.g. after the first World War for the U.K., during the Great Depression for Canada and the United States, during World War II for France, Germany, Italy and Norway); (ii) a change in the rate of growth (e.g. after 1940 for Australia, after the Great Depression for Canada, after World War II for France, Germany, Italy and Norway). A feature of interest is that major changes in levels and/or slopes seem to occur only once in the sample considered. In particular when changes in both the level and slope occur, they seem to do so roughly at the same time (e.g. France, Germany and Italy). This is not to say that other changes in level and/or slope do not occur for some series but rather that if they do their magnitude are small compared to the changes described previously. This observation supports our claim that the restricted case for which the trend function changes only once appears to be of substantial relevance in practice.

A second feature of interest is that the transition path to a new trend appears to occur gradually, especially when a change in level is involved. This suggests that when using Models (1) and (2), the innovational outlier version would be more appropriate. After visual inspection of the data, it was decided to apply the following models to each country: Innovational Outlier Model (1) for the United Kingdom and the United States; Innovational Outlier Model (2) for Canada, Denmark, Finland, France, Germany, Italy and Sweden; and Additive Outlier Model (3) for Australia and Norway.

Some comments are in order to offer guidance as to the appropriate choice of models in general. Note first that Model (2) is the most general and allow, as possible subsets, changes described in Models (1) and (3). Model (2) could therefore be used for all series. The problem with this approach is that it may lead to tests with lower power compared to what could be obtained using a more constrained model. As will be seen in Section 4.5, the asymptotic critical values are the highest under Model (2). There is therefore an advantage to select a model containing no irrelevant regressors, for example in selecting Model (1) over Model (2) if no change in slope is apparent and, similarly, Model (3) over Model (2) when a change in slope is apparent but not a change in level. However, care must be exercised not to select a model that is too constrained since this would imply a substantial loss in power and even tests that are potentially inconsistent. When in doubt, one should start with Model (2) since it is more general. However, a good practical rule is to assess the robustness of non-rejections of the unit root to different model selections.

The dotted line in each graph represents the fitted trend function. Each non-linear trend function is obtained using the

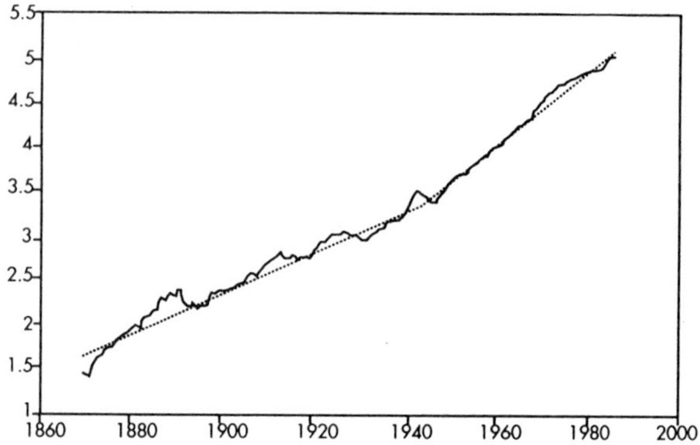

Figure 4.1.A Australia

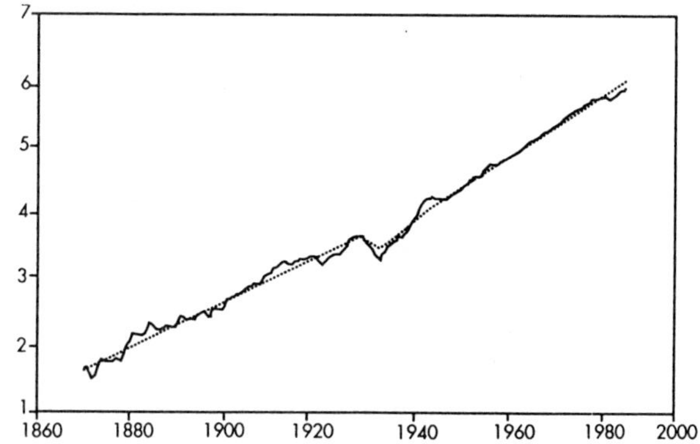

Figure 4.1.B Canada

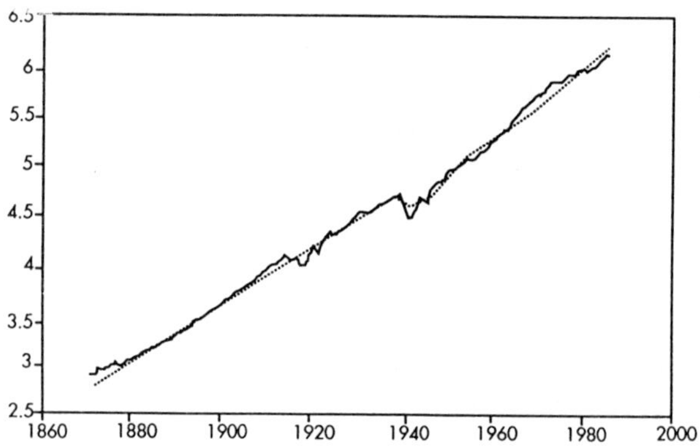

Figure 4.1.C Denmark

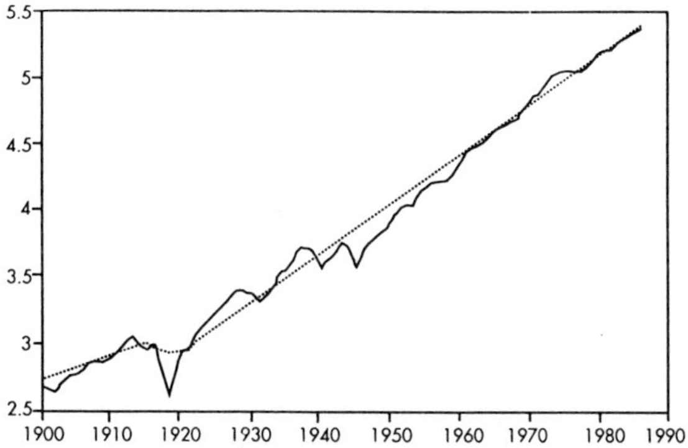

Figure 4.1.D Finland

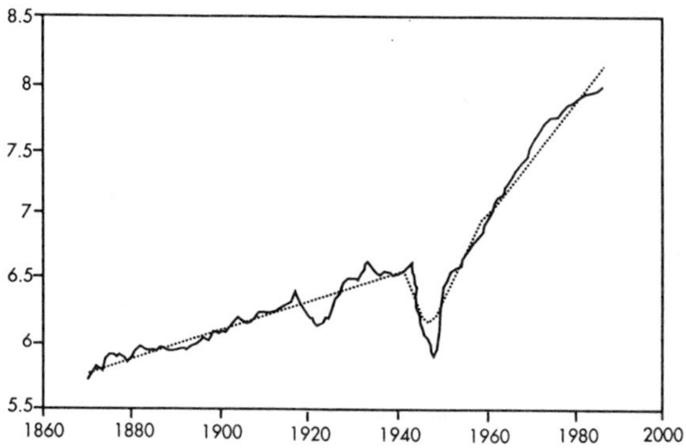

Figure 4.1.E France

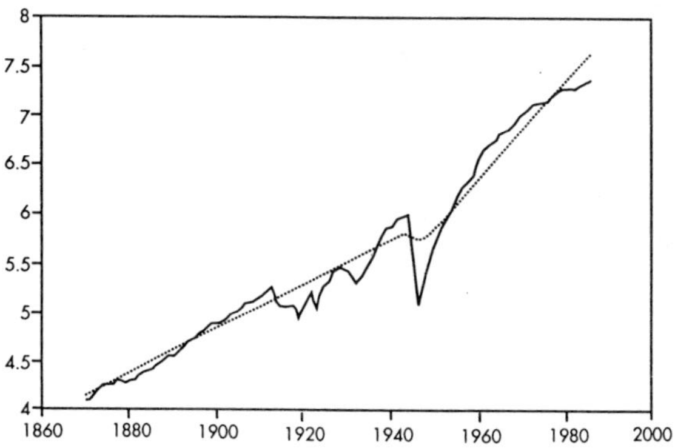

Figure 4.1.F Germany

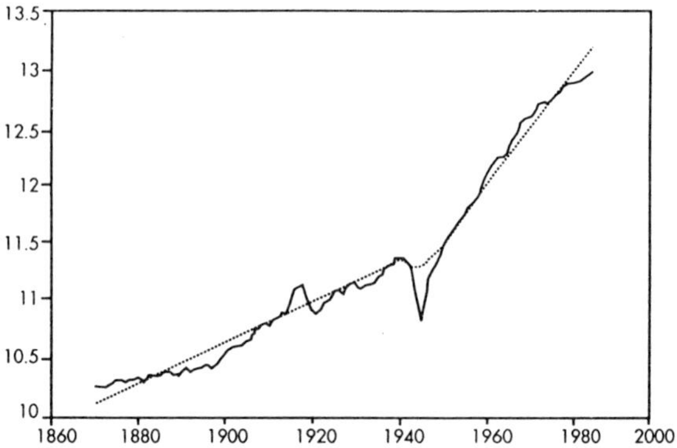

Figure 4.1.G Italy

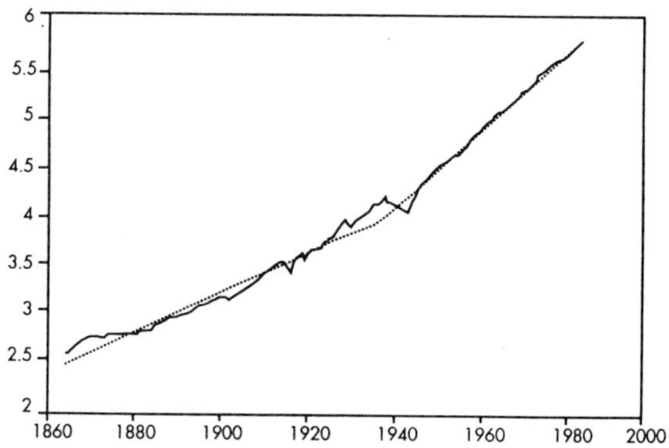

Figure 4.1.H Norway

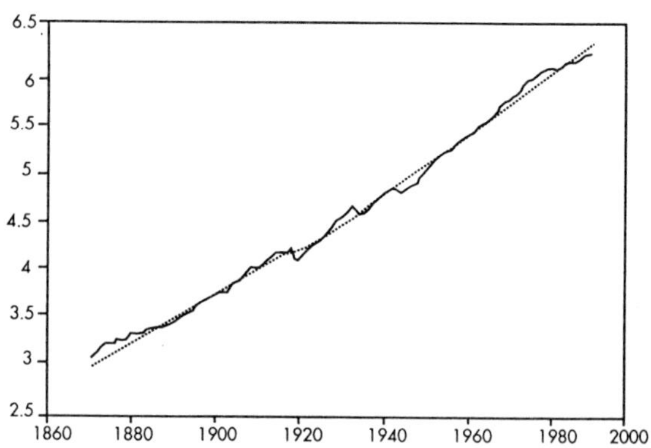

Figure 4.1.I Sweden

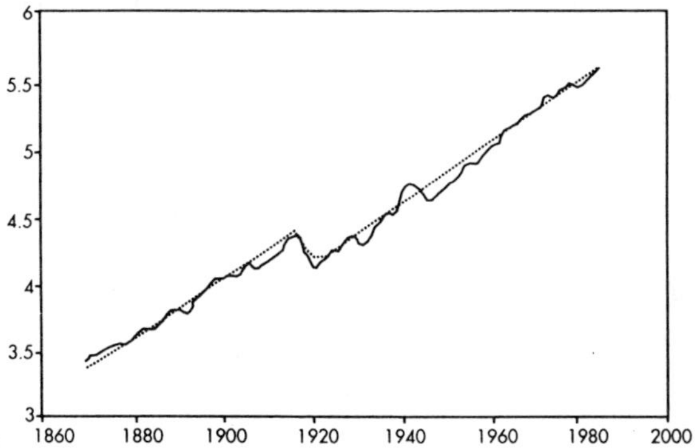

Figure 4.1.J United Kingdom

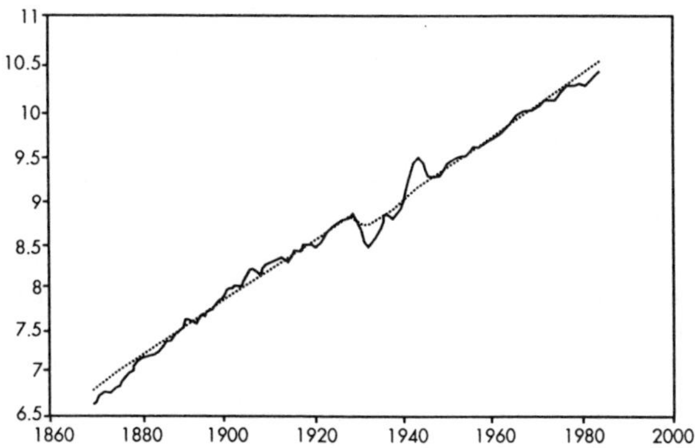

Figure 4.1.K United States

appropriate model under the alternative hypothesis. As stated, only one break date was allowed and is as stated in each graph for the different countries. We postpone until Section 4.6, the precise method under which these break dates were selected. For Australia and Norway the trend functions show a kink at the time of the break since the model selected for these countries is the additive outlier version of Model (3). In these cases, the trend function is simply the fitted values from a least-squares regression applied to equation (3).

For the other countries, an innovational outlier model was preferred which translates into a nonlinear trend function showing a gradual adjustment to the new path following the break date. As can be seen, the adjustment typically lasts only a few years and seems to roughly capture the effects that occur following a "crash". The fitted trend functions are more cumbersome to compute than in the case of the additive outlier model. The precise method can be described as follows. First the following regression is estimated by *OLS* in the case of Model (2) (the case of Model (1) is similar without the regressor DT_t^*):

$$y_t = c + bt + \theta DU_t + \gamma DT_t^* + \sum_{i=1}^{p} a_i y_{t-i} + e_t. \tag{8}$$

The choice of the truncation lag for the autoregressive order is selected endogenously in the same procedure that selects the break point T_b and we postpone its discussion to Section 4.6. The next step is to use the least-squares estimates $(\hat{c}, \hat{b}, \hat{a}_i)$ to obtain estimates of the coefficients (μ, β, Φ_i) in equation (5). This is achieved by noting the following relations that can be obtained expressing equation (5) in the form of regression (8). First let $(1 - \sum_{i=1}^{p} \hat{a}_i L^i) = \hat{A}(L)$. Since we are approximating the general *ARMA* process for the noise component by a finite autoregression, we have the relation $\hat{A}(L) = \hat{\Phi}(L)^{-1}$. We also define $\hat{A}(1) = (1 - \sum_{i=1}^{p} \hat{a}_i)$. Then, we have $\hat{\beta} = \hat{b}/\hat{A}(1)$, $\hat{\mu} = (\hat{c} - \hat{b}\hat{\Upsilon})/\hat{A}(1)$ where $\hat{\Upsilon} = \sum_{i=1}^{p} i\hat{a}_i$ is the so-called mean lag matrix. The fitted trend function, call it $TR(2)_t$, say, for Model (2), is then given by:

$$TR(2)_t = \hat{\mu} + \hat{\beta}t + \hat{A}(L)^{-1}(\hat{\theta} DU_t + \hat{\gamma} DT_t^*),$$

which can be solved recursively. The method described above is a relatively easy and flexible way of obtaining fitted trend functions that allow a gradual adjustment following a break. As seen from the graphs it also fits the general pattern of the data adequately.

As a preliminary investigation about the possible stationarity of the noise component, we considered the autocorrelation function of the estimated cyclical component, namely the difference between the actual series and the fitted trend computed as described above. The results are present in Table 4.1.

The first line in Table 4.1 reports the first six estimated autocorrelations when a linear time trend is assumed, i.e. the autocorrelations of the residuals from a least-squares regression of the series on a constant and a time trend. The results parallel those

Table 4.1

Sample Autocorrelations of the Detrended Series

Series	*Trend*	r_1	r_2	r_3	r_4	r_5	r_6
Australia	*Linear*	.96	.91	.85	.80	.75	.71
	AO-3 (1943)	.83	.66	.46	.30	.16	.07
Canada	*Linear*	.92	.82	.75	.67	.60	.54
	IO-2 (1929)	.75	.53	.34	.17	.03	−.03
Denmark	*Linear*	.91	.81	.75	.67	.58	.51
	IO-2 (1938)	.73	.51	.44	.31	.12	−.02
Finland	*Linear*	.88	.73	.64	.55	.45	.37
	IO-2 (1915)	.80	.55	.44	.33	.22	.11
France	*Linear*	.96	.90	.84	.77	.70	.64
	IO-2 (1941)	.78	.51	.29	.10	−.06	−.12
Germany	*Linear*	.93	.82	.73	.64	.55	.48
	IO-2 (1943)	.84	.62	.46	.32	.18	.07
Italy	*Linear*	.95	.90	.84	.78	.73	.69
	IO-2 (1941)	.83	.62	.45	.33	.23	.18
Norway	*Linear*	.95	.90	.85	.80	.74	.69
	AO-3 (1941)	.88	.75	.65	.52	.39	.30
Sweden	*Linear*	.94	.86	.78	.70	.61	.55
	IO-2 (1915)	.86	.70	.55	.40	.25	.14
U.K.	*Linear*	.94	.84	.73	.64	.58	.53
	IO-1 (1917)	.84	.62	.40	.23	.13	.04
U.S.	*Linear*	.88	.70	.51	.35	.24	.19
	IO-1 (1928)	.80	.55	.32	.15	.07	.08

Notes:

For each country, the first line presents the first six estimated autocorrelation coefficients of the linearly detrended series, i.e. of the residuals y_t^* in the regression $y_t = \mu + \beta t + y_t^*$ estimated by *OLS*. The second line gives similar results with trends allowed to be non-linear. *AO* and *IO* denote the additive and innovational outlier models, respectively. The digit next to the *AO* or *IO* characterization indicates the model considered. The date in parenthesis is the date of the break specified (see Section 4.6 for a discussion of the method used to select these break dates).

of Nelson and Plosser (1982) in that the autocorrelations are high and decay slowly. As argued by Nelson and Plosser (1982), this is the pattern one would expect if the actual series were processes with a unit root. The second lines in Table 4.1 present the autocorrelations of the cyclical components estimated using one of the non-linear trend functions described above, i.e. the difference between the actual series and the fitted trend function allowing a single break. The results are dramatically different and show a rapid decay of the autocorrelation coefficients. This is particularly the case for Australia, Canada, Denmark, France, Germany, the United Kingdom and the United States. These estimates show a pattern that one would expect if the noise component was indeed stationary, i.e. without a unit root. This analysis, however, cannot provide a formal test of the unit root hypothesis versus the alternative hypothesis of stationary fluctuations and should simply be viewed as preliminary evidence about the possibility that many series may be characterized by stationary fluctuations around a trend function with a one-time change in level and/or slope. Formal statistical procedures are described in Section 4.5.

4.4 THE EFFECT OF BREAKS ON STANDARD UNIT ROOT TESTS

In this section, we briefly discuss the effect of using standard unit root tests that do not allow for a possible change when indeed such a change occurs. Among the many tests for unit roots available the most widely used is the one proposed by Dickey and Fuller (1979) and extended by Said and Dickey (1984); see, e.g. the reviews by Campbell and Perron (1991) and Stock (1992). In the case of trending data, this test is based on the t-statistic for testing $\alpha = 1$, denoted $t_{\tilde{\alpha}}$, in the following regression:

$$y_t = \mu + \beta t + \alpha y_{t-1} + \sum_{i=1}^{k} a_i \Delta y_{t-i} + e_i. \tag{9}$$

In the case where the noise component is an autoregressive process of order p, the truncation lag parameter k should be selected at least as great as p, in which case the limiting distribution of $t_{\tilde{\alpha}}$ is tabulated in Fuller (1976) and is substantially different from the standard normal distribution. In the case where the noise component of the first-difference of y_t, under the null hypothesis is a more general *ARMA* process, the trick is to recognize that we can use an autoregressive approximation. Said and Dickey (1984) show under what conditions this approximation leads to the same asymptotic distribution as tabulated in Fuller (1976). In practice, the choice of the truncation lag is a difficult issue and we postpone a discussion of various methods to Section 4.5.

To show the effect of the presence of a break in the trend function on the behaviour of the test statistic $t_{\tilde{\alpha}}$ we considered the following simulation experiment. Two possible data generating processes involving, first, a series with a change in level, and secondly, a case where the series show a change in slope are considered. To be more precise the data-generating processes considered are:

$$y_t = \mu_2 DU_t + e_t \tag{10}$$

and

$$y_t = \beta_1 + (\beta_2 - \beta_1)DT_t^* + e_t, \tag{11}$$

for $t = 1, \ldots, T$, where $e_t \sim IID.N(0,1), T = 100$ and $T_b = 50$ (i.e. a shift occurring at mid-sample). In equation (10) the process has mean 0 up to time T_b and μ_2 afterwards. In equation (11) the process has slope $\beta_1 = 1$ up to time T_b and slope $(\beta_1 - \beta_2)$ afterwards. Note that the noise component here is an *IID*. sequence. It is therefore an extreme case where the unit root hypothesis is clearly violated. We studied the behaviour of $t_{\tilde{\alpha}}$ computed from regression (9) with $k = 0, 2$, and (11) using 5,000 replications generated from either (10) or (11).

Figure 4.2 presents the results for the case of a change in mean where the data are generated from (10) with $\mu_2 = 0, 2, 5, 10$ and 25 (0 being the base case of no change). The graphs present the cumulative distribution functions (*cdf*) of the t-statistic $t_{\tilde{\alpha}}$. As can be seen from the graphs, the *cdf* of $t_{\tilde{\alpha}}$ is substantially shifted to the right as the change in mean increases. For $k = 0$, the test has little power when the change rises to 10 and none at all when the change is as high as 25 (noting that the rejection region are values of $t_{\tilde{\alpha}}$ less than -3.41 at the 5% level; see Fuller (1976). These values may seem high but it is important to notice that we are considering the extreme case of a noise component that is *IID*.. With a noise component positively correlated, as is the case in practice, the test no longer rejects for much lower values of the change in mean. Panels (B) and (C) show not only

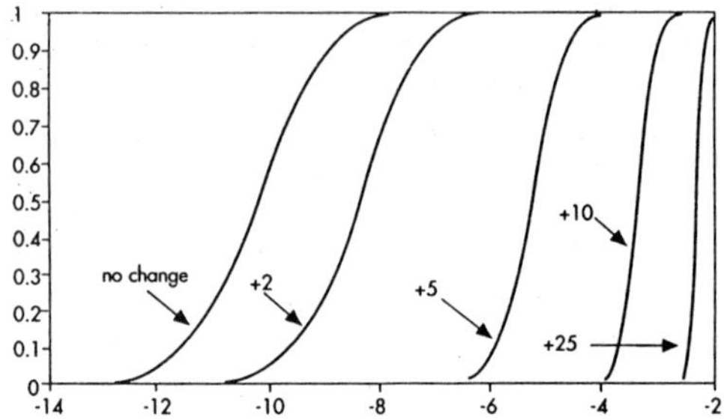

Figure 4.2.A Change in intercept: k = 0.

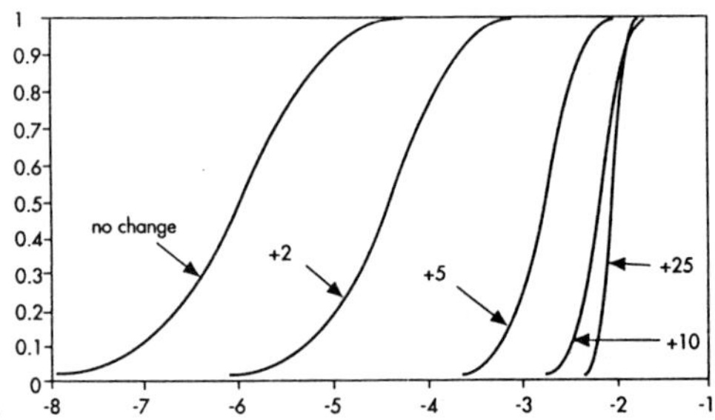

Figure 4.2.B Change in intercep: k = 2

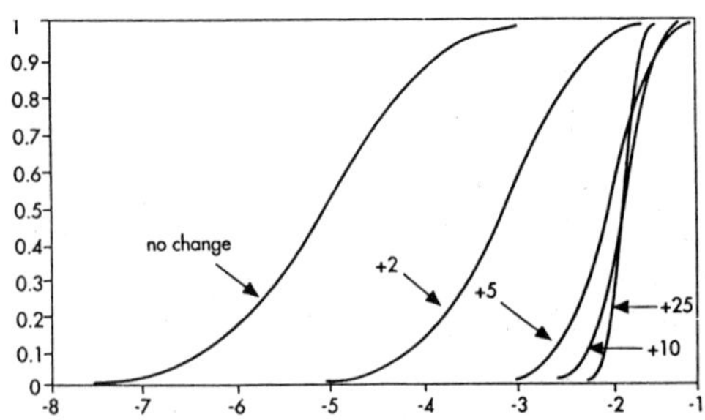

Figure 4.2.C Change in intercept: k = 5

Notes:

Cumulative distribution function of $t(\tilde{\alpha})$, the t-statistic for $\alpha = 1$ in the regression $y_t = \mu + \beta t + \alpha y_{t-1} + \sum_{i=1,k} c_i \Delta y_{t-i} + e_t$(estimated by OLS) when the data-generating process is $y_t = \mu_2 DU_t + e_t$ ($e_t \sim IID\ N(0,1)$ and $DU_t = 1$ if $t > T_b$ and 0 otherwise). Simulated values are based on 5,000 replications. The entries next to each curve are the different values of μ_2, the change in mean.

that this feature of reduced power remains as the autoregressive truncation lag increases but also that it becomes more severe. For $k = 5$, the test no longer has any power when the change is 5.

Figure 4.3 presents the results for the case where the data-generating process involves a change in slope and is generated from equation (11) with $\beta_1 = 1$ and the change $\beta_2 - \beta_1 = 0$ (the base case), $-.1, -.3, -.6$ and $-.8$. The same qualitative features emerge, namely substantially reduced power as the change in slope increases, more so as the truncation lag k increases. Of interest is the fact that the test looses all power for realistic values of the change in slope even if the data-generating process involves the extreme case of an *IID*. sequence. For example, with $k = 2$ the test has no longer any power when the change in slope is greater than 0.3 (in absolute value) and when $k = 5$, the corresponding value of 0.1.

The simulation results described above show that the non-rejection of the unit root hypothesis frequently reported in practice is consistent with the possibility that the processes be characterized by stationary fluctuations around a trend function with a one time change in level and/or slope. The next section discusses a different class of tests that avoids these reduced power problems and allow for the possibility of a change in the trend function.

4.5 THE TEST PROCEDURES

The test procedures are based on simple autoregressions (estimated by *OLS*) that are appropriately augmented with trend and dummy components. The test statistics are based on the values of the t-statistics for testing that the sum of the autoregressive coefficients is equal to one. The procedures follow the approach of Said and Dickey (1984) by approximating *ARMA* processes by autoregressions of order k. It is important to note that the particular regression framework is different for each model and also different across the innovational and additive outlier versions. The asymptotic critical values also differ according to the specific framework considered.

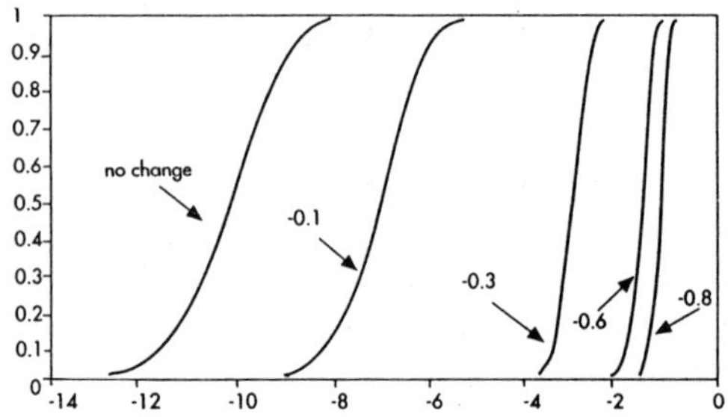

Figure 4.3.A Change in slope: k = 0.

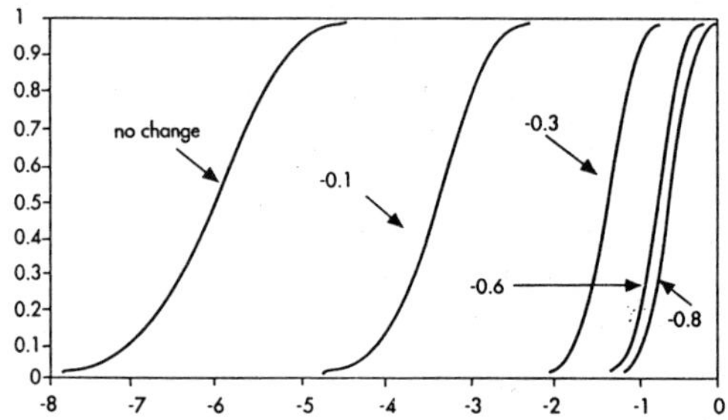

Figure 4.3.B Change in slope: k = 2

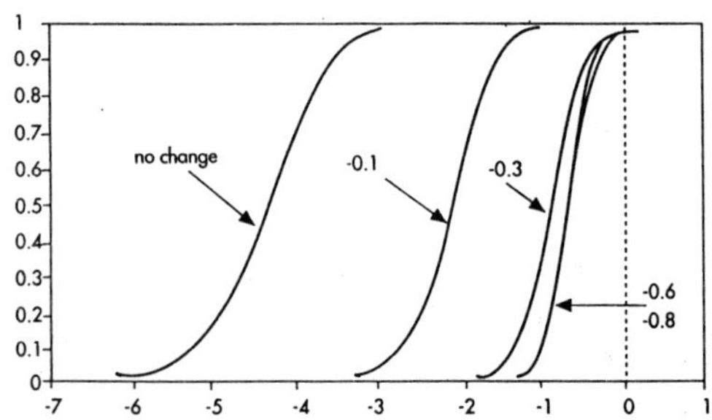

Figure 4.3.C Change in slope: k = 5

Notes:

Cumulative distribution function of $t(\tilde{\alpha})$, the t-statistic for $\alpha = 1$ in the regression $y_t = \mu + \beta t + \alpha y_{t-1} + \sum_{i=1,k} c_i \Delta y_{t-i} + e_t$(estimated by OLS) when the data-generating process is $y_t = \beta_1 t + (\beta_2 - \beta_1) DT_t^* + e_t(\beta_1 = 1$, $e_t \sim IID\ N(0,1)$ and $DT_t^* = (t - T_b)$ if $t > T_b$ and 0 otherwise). Simulated values are based on 5,000 replications. The entries next to each curve are the different values of the change in the slope $\beta_2 - \beta_1$.

4.5.1 The Innovational Outlier Models

We start with the innovational outlier version of Models (1) and (2). The regressions, at the basis of the testing procedures, are as follows for Models (1) and (2), respectively:

$$y_t = \mu + \beta t + \gamma DU_t + \delta D(T_b)_t + \alpha y_{t-1} + \sum_{i=1}^{k} a_i \Delta y_{t-i} + e_t \tag{12}$$

and

$$y_t = \mu + \beta t + \theta DU_t + \gamma DT_t^* + \delta D(T_b)_t + \alpha y_{t-1} + \sum_{i=1}^{k} a_i \Delta y_{t-i} + e_t \tag{13}$$

where we recall that $DU_t = 1$ and $DT_t^* = (t - T_b)$ if $t > T_b$ (0 otherwise), and $D(T_b)_t = 1$ if $t = T_b + 1$ (0 otherwise). The statistic of interest is the t-statistic for testing that $\alpha = 1$. The dummy variable $D(T_b)_t$ is introduced following Perron (1989). In the case where the break point T_b is treated as known, it allows the t-statistic for testing $\alpha = 1$ to be invariant in finite samples to the value of the change in the intercept under the null hypothesis of a unit root. When T_b is unknown, such a finite sample invariance property no longer holds although the statistics are asymptotically invariant.[2] In regressions (12) and (13) there are two parameters that may be unknown in practice, namely k and T_b. We discuss below various data-dependent methods to select them and the ensuing effects on the construction of the tests and their limiting distributions.

4.5.2 The Additive Outlier Models

For the additive outlier models, the procedures are different and consist of two-step approaches. In the first step, the trend function of the series is estimated and removed from the original

---

2. Note that Zivot and Andrews (1992) do not introduce the additional one-time dummy $D(T_b)_t$ in the regression they consider for the innovational outlier Models (1) and (2). Since this regressor is asymptotically negligible, this makes little difference for moderate to large samples in the case where the break point is treated as unknown. We prefer to keep this dummy regressor for all methods of choosing the date of break (see below) since it naturally extends the case of the known break point where it allows invariance to the change in intercept under the null hypothesis.

series via the following regressions for Models (1), (2) and (3), respectively:

$$y_t = \mu + \beta t + \gamma DU_t + \tilde{y}_t \tag{14}$$

$$y_t = \mu + \beta t + \theta DU_t + \gamma DT_t^* + \tilde{y}_t \tag{15}$$

$$y_t = \mu + \beta t + \gamma DT_t^* + \tilde{y}_t \tag{16}$$

where $\tilde{y}_t$ is accordingly defined as the detrended series. The next step differs according to whether or not the first step involves DU_t, the dummy associated with a change in intercept.[3] For Models (1) and (2), the test is based on the value of the t-statistic for testing that the sum of the autoregressive coefficients is equal to 1 ($\alpha = 1$) in the following autoregression applied to the estimated noise component $\tilde{y}_t$:

$$\tilde{y}_t = \alpha \tilde{y}_{t-1} + \sum_{j=0}^{k} d_j D(T_b)_{t-j} + \sum_{i=1}^{k} a_i \Delta \tilde{y}_{t-i} + e_t. \tag{17}$$

Details about the need to introduce the current value and lags of the dummies $D(T_b)_t$ can be found in Perron and Vogelsang (1992c). For Model (3) where no change in level is involved and the two segments of the trend are joined at the time of break, there is no need to introduce the dummies in the second step regression. Hence, for Model (3), the second step regression is of the form:

$$\tilde{y}_t = \alpha \tilde{y}_{t-1} + \sum_{i=1}^{k} a_i \Delta \tilde{y}_{t-i} + e_t. \tag{18}$$

This two-step procedure allows a test for a unit root that is asymptotically invariant to γ (the magnitude of the change in slope) under the null hypothesis.[4]

---

3. Note that the following discussion about the appropriate two-step procedures is different than that stated in Perron (1989). This paper contained an error in the treatment of the additive outlier models which is corrected in Perron (1993) and discussed more extensively in Perron and Vogelsang (1992c). For the case of non-trending data see Perron (1990a) and Perron and Vogelsang (1992b).

4. Note that such an asymptotic invariance no longer holds in the one-step procedure considered by Banerjee, Lumsdaine and Stock (1992) and Zivot and Andrews (1992). They consider a regression similar to (12) and (13) associated with the innovational outlier Models (1) and (2) of the form

The construction of the test statistics of interest, namely the t-statistic for testing $\alpha = 1$ in equations (12) to (13) or (15) to (17) depend on two parameters that are in general unknown: the break that T_b and the truncation lag parameter k. We now discuss several ways to choose them and the corresponding appropriate critical values. Note at the outset, however, that only the various ways of choosing the break date influence the appropriate asymptotic critical values to be used, not the particular ways of choosing the truncation lag parameter.

4.5.3 Methods to Choose the Break Date

In some instances, the break date T_b may be treated as known or one may wish to condition the inference on a particular given value. In that case, one therefore needs to estimate the basic autoregressions for only one value of T_b, that which is specified *a priori*. The critical values to be used, however, depend on the particular model selected. For Models (1) and (2) the critical values can be found in Perron (1989, Tables IV.B and VI.B) and are the same in the innovational or additive outlier versions. For the additive outlier Model (3), the critical values can be found in Perron (1993). Note that the critical values depend on the relative position of the break in the sample. However, unless the break occurs near the beginning or the end of the sample, there is not much variation in the critical values. For this reason we only report, in Tables 4.2 to 4.6 for the various models, those corresponding to a break at mid-sample (denoted $\lambda = 0.5$, with λ the ratio T_b/T).

$y_t = \mu+\beta t+\gamma DT_t^* +\alpha y_{t-1}+\sum_{i=1,k} a_i y_{t-i} +e_t$. The asymptotic distribution of the associated t-statistic for $\alpha = 1$ is not invariant to the magnitude of the change in slope under the null hypothesis. Indeed, if there is such a change in slope, the test will always reject in large samples irrespective of the presence or absence of a unit root in the noise component. Hence, the test analyzed by these authors is a joint test of the null of a unit root and no change in the slope and not a test of the unit root hypothesis per se where the change in slope is allowed under both the null and alternative hypotheses. Note that even in the case where the change in slope is zero, the asymptotic distribution is different from that of the t-statistic associated with the two-step regressions (16) to (18).

We now consider two methods to select T_b endogenously. As in Zivot and Andrews (1992), Banerjee, Lumsdaine and Stock (1992) and Perron (1990), we first consider the procedure whereby T_b is selected as the value, over all possible break points, which minimizes the t-statistic for testing $\alpha = 1$ in the appropriate autoregression. The investigator therefore needs to estimate the autoregressions for all possible values of the break date T_b.[5] The critical values for the innovational outlier Models (1) and (2) can be found in Zivot and Andrews (1992). For the additive outlier version, they can be found in Perron (1990) for Model (3) and Perron and Vogelsang (1993) for Models (1) and (2).[6] They are reproduced in Tables 4.2 to 4.6 in the lines $t(\alpha)$.

An alternative method to select the break point is, following Christiano (1992), to consider choosing T_b as the value, over all possible break points, which maximizes (or minimizes) the value of the t-statistic for testing $\gamma = 0$ in regressions (12) to (16). This procedure imposes the mild *a priori* restriction of a one-sided change in the trend function. In the case of Model (1), since only sudden crashes are of interest for the various series described in Section 4.2, the procedure is to choose T_b as that value which minimizes the t-statistic on the coefficient associated with the change in intercept DU_t. In the case of Models (2) and (3), since only more or less sudden increases in slope are observed, the interest is in choosing T_b as the value which maximizes the t-statistic on the coefficient associated with DT_t^*, the change in slope. For all cases, one can avoid having to make the assumption of a one-sided change by maximizing the absolute value of the t-statistic for testing $\gamma = 0$. For Models (1) and (3), this leads to a procedure that has properties similar to that whereby T_b is chosen by minimizing the t-statistic for testing $\alpha = 1$. For Model (2), it can lead to a substantial increase in power; see Perron and

5. Banerjee, Lumsdaine and Stock (1992) and Zivot and Andrews (1992) suggest estimating the autoregressions for all values of the break point in some interval that excludes break dates near the beginning or the end of the sample, e.g. for all values of T_b in the interval $(.15T, .85T)$. This makes little difference in theory or in practice.

6. Note that these papers and others that we cite below also report extensive finite sample critical values obtained via simulations with and without data-dependent methods to select the truncation lag parameter k (see section 4.5.4 below).

Table 4.2
Asymptotic Distribution: Innovational Outlier Model (1)

	1.0%	2.5%	5.0%	10.0%	90.0%	95.0%	97.5%	99.0%
$\lambda = 0.5$	−4.32	−4.01	−3.76	−3.46	−1.17	−0.79	−0.49	−0.15
$t(\alpha)$	−5.34	−5.02	−4.80	−4.58	−2.99	−2.77	−2.56	−2.32
$t(\gamma)$	−5.15	−4.87	−4.64	−4.38	−2.27	−1.85	−1.38	−0.70
$t(\|\gamma\|)$	−5.33	−5.08	−4.84	−4.59	−2.71	−2.35	−2.01	−1.54

Table 4.3
Asymptotic Distribution: Innovational Outlier Model (2)

	1.0%	2.5%	5.0%	10.0%	90.0%	95.0%	97.5%	99.0%
$\lambda = 0.5$	−4.90	−4.53	−4.24	−3.96	−1.96	−1.69	−1.43	−1.07
$t(\alpha)$	−5.57	−5.30	−5.08	−4.82	−3.25	−3.06	−2.91	−2.72
$t(\gamma)$	−5.28	−4.94	−4.62	−4.28	−1.64	−1.33	−0.98	−0.59
$t(\|\gamma\|)$	−5.57	−5.20	−4.91	−4.59	−2.15	−1.86	−1.59	−1.30

Table 4.4
Asymptotic Distribution: Additive Outlier Model (1)

	1.0%	2.5%	5.0%	10.0%	90.0%	95.0%	97.5%	99.0%
$\lambda = 0.5$	−4.32	−4.01	−3.76	−3.46	−1.17	−0.79	−0.49	−0.15
$t(\alpha)$	−5.34	−5.02	−4.80	−4.58	−2.99	−2.77	−2.56	−2.32
$t(\gamma)$	−4.57	−4.24	−4.01	−3.74	−1.85	−1.57	−1.34	−1.02
$t(\|\gamma\|)$	−4.70	−4.40	−4.17	−3.90	−2.05	−1.78	−1.52	−1.22

Table 4.5
Asymptotic Distribution: Additive Outlier Model (2)

	1.0%	2.5%	5.0%	10.0%	90.0%	95.0%	97.5%	99.0%
$\lambda = 0.5$	−4.90	−4.53	−4.24	−3.96	−1.96	−1.69	−1.43	−1.07
$t(\alpha)$	−5.57	−5.30	−5.08	−4.82	−3.25	−3.06	−2.91	−2.72
$t(\gamma)$	−4.82	−4.54	−4.18	−3.84	−1.34	−0.96	−0.63	−0.21
$t(\|\gamma\|)$	−5.01	−4.71	−4.43	−4.13	−2.11	−1.79	−1.54	−1.18

Vogelsang (1993). If the one-sided change condition is imposed *a priori*, the resulting procedure allows greater power. Note that in practice, for the estimation of the statistic, it does not matter if the one-sided condition is imposed or not. This difference only affects the interpretation of the significance of the estimates.

The critical values appropriate to these methods of choosing the break date are derived in Perron and Vogelsand (1993) and are reproduced in Tables 4.2 to 4.6. For the case where the one-sided change is imposed a priori the critical values are those in

Table 4.6
Asymptotic Distribution: Additive Outlier Model (3)

	1.0%	2.5%	5.0%	10.0%	90.0%	95.0%	97.5%	99.0%
$\lambda = 0.5$	−4.49	−4.17	−3.93	−3.65	−1.80	−1.47	−1.21	−0.85
$t(\alpha)$	−4.91	−4.60	−4.36	−4.09	−2.32	−2.12	−1.97	−1.78
$t(\gamma)$	−4.67	−4.36	−4.08	−3.77	−1.57	−1.22	−0.90	−0.49
$t(\|\gamma\|)$	−4.87	−4.58	−4.34	−4.04	−2.14	−1.87	−1.61	−1.30

Notes:
The sources for the asymptotic critical values are as follows:
(i) fixed $\lambda = 0.5$, Perron (1989); (ii) $t(\alpha)$, Zivot and Andrews (1992) for Tables 4.2 and 4.3, Perron and Vogelsang (1993) for Tables 4.4 and 4.5, and Perron (1990) for Table 4.6; (iii) $t(\gamma)$ and $t(|\gamma|)$, Perron and Vogelsang (1993).

the lines with the heading $t(\gamma)$. Those for the case where no such condition is imposed are in the lines with the heading $t(|\gamma|)$.

4.5.4 Methods to Select the Truncation Lag Parameter k

There is now substantial evidence that using data-dependent methods to select the truncation lag parameter k leads to test statistics having better properties (stable size and higher power) than if a fixed k is chosen *a priori* unless, of course, one happens to select that value of k which is best; see Ng and Perron (1993) and Perron and Vogelsang (1992a). We consider, in the empirical applications below, two such data-dependent methods. The first is the one originally implemented by Perron (1989). It uses a general to specific recursive procedure based on the value of the t-statistic on the coefficient associated with the last lag in the estimated autoregression. More specifically, the procedure selects that value of k, say k^*, such that the coefficient on the last lag in an autoregression of order k^* is significant and that the coefficient on the last lag in an autoregression of order greater than k^* is insignificant, up to some maximum order $kmax$ selected *a priori*. In the empirical applications reported below, we use a two-sided 10% test based on the asymptotic normal distribution to assess the significance of the last lags and $kmax$ is set to 10. An alternative procedure to select k is to use an information criterion. In our empirical applications, we report results whereby

k is chosen to minimize the Akaike Information Criterion (AIC).[7]

In the case where the noise function is assumed to be generated from a finite order pure autoregressive processes, we can use results in Hall (1991) to show that the two data-dependent methods described above lead to tests having the same asymptotic distribution as would prevail if the true autoregressive order was selected to estimate the autoregression. In the case of the recursive t-statistic procedure, this holds provided $kmax$ is chosen greater than the true value. Things are more complex if we allow the (more realistic) possibility of moving-average components in the data-generating process. This issue is investigated in Ng and Perron (1993). Our results indicate that both procedures lead to the same asymptotic distribution as in the fixed k case provided $kmax$ increases to infinity as the sample size (T) increases in a way such that $(Kmax)^3/T$ converges to 0. However, the finite sample performances appear rather different with a moving-average component. Data-dependent methods based on information criteria, such as the AIC, tend to select very parsimonious models leading to tests with sometimes serious size distortions. This finite sample performance is consistent with our finding that the use of an information criterion leads to a selected value of k that increases to infinity, as T increases, only at rate $log(T)$, a very slow rate. These theoretical results are in accord with the empirical results reported below for which the procedures based on the AIC lead to very small values of k being selected (typically 0 or 1). This suspicion about the performance of data-based methods using the AIC is reinforced by the fact that often the estimated residuals exhibit serial correlation. These instances are indicated in the reported results. For these reasons, we put more confidence on the results obtained using

7. Among both classes considered for the selection of the truncation lag parameter k, there are many other variants that could be used. For example, among procedures based on information criteria, another example is the Bayesian Information Criterion (BIC). Since, BIC generally yields more parsimonious models than AIC, the same qualitative conclusions discussed below remain. Among the class of procedures based on tests of significance on the coefficients of the lags, one could also use recursive joint tests assessing the significance of several coefficients on the last lags of the autoregressions. Such procedures yields qualitatively similar results as the recursive t-statistic discussed here.

the recursive t-statistic procedure.

In the empirical results reported below we use the following notation:

(i) $t^{\alpha}(t)$, the value of the t-statistic for testing $\alpha = 1$ when T_b is chosen to minimize this t-statistic with k chosen according to the recursive t-statistic on the coefficient of the last lag;

(ii) $t^{\alpha}(aic)$, same as (i) except that k is chosen according to the *AIC*;

(iii) $T^{\gamma}(t)$, the value of the t-statistic for testing $\alpha = 1$ when T_b is chosen to maximize (or minimize) the t-statistic on γ (the coefficient on the change in slope or intercept) with k chosen according to the recursive t-statistic on the coefficient of the last lag;

(iv) $t^{\gamma}(aic)$, the same as (iii) except that k is chosen according to the *AIC*.

4.6 EMPIRICAL APPLICATIONS

Table 4.7 presents the empirical results for the historical *GDP* series for each of the 11 countries with the corresponding model of the selected trend function. Columns 1 and 2 give, respectively, the date of break in the trend function and the value of the truncation lag parameter k in the autoregression being selected by the various methods. Columns 3, 4 and 5 present key estimated parameters of the autoregressions along with their t-statistics in parentheses: $\hat{\beta}$ is the estimate of the initial (pre-break) slope of the trend function, $\hat{\theta}$ is the estimate of the change in the intercept of the trend function in the case of Model (2), and $\hat{\gamma}$ is the estimate of the change in the slope of the trend function for Models (2) and (3). For Model (1) it is the estimate of the change in the intercept of the trend function. Column 6 and 7 present the key estimates related to the testing procedure described in the last section, namely the estimate of the sum of the autoregressive coefficient, $\hat{\alpha}$, and its associated t-statistic for testing that $\alpha = 1$, $t_{\hat{\alpha}}$. The last four columns from 8 to 11 present p-values associated with the four different tests described earlier. As mentioned before, these p-values are based on the corresponding asymptotic distribution of the test statistics under the null hypothesis of a unit root in the noise component (allowing a break or not).

Table 4.7

Empirical Results: Real GDP Series

							p-values			
T_b	k	$\hat{B}$	$\hat{\theta}$	$\hat{\gamma}$	$\hat{\alpha}$	$t_{\hat{\alpha}}$	$t^{\alpha}(t)$	$t^{\alpha}(aic)$	$t^{\gamma}(t)$	$t^{\gamma}(aic)$
(1)	(2)	(3)	(4)	(5)	(6)	(7)	(8)	(9)	(10)	(11)
				Australia: Model (3)						
1943	2	.022		.018	.782	−4.52	.03			
		(52.89)		(17.08)						
1939	0	.022		.018	.848	−3.52		.29		
		(47.04)		(16.43)						
1947	2	.023		.020	.791	−4.29			.03	
		(58.71)		(17.55)					(.06)	
1947	1	.023		.020	.844	−3.44				.18
		(58.71)		(17.55)						(.31)
				Canada: Model (2)						
1929	4	.013	−.117	.005	.615	−5.26	.04		.01	
		(5.19)	(−3.84)	(4.50)					(.02)	
1928	0	0.08	−.300	.003	.738	−4.43		.27*		.07*
		(5.36)	(−3.84)	(3.57)						(.14*)
				Denmark: Model (2)						
1938	7	.013	−.421	.004	.510	−5.94	<.01		<.01	
		(5.91)	(−5.56)	(5.42)					(<.01)	
1938	0	.008	−.282	.003	.703	−4.98		.08*		.02*
		(5.00)	(−4.39)	(4.21)						(0.4*)
				Finland: Model (2)						
1915	3	.005	−.327	.005	.697	−4.08	.47			
		(1.18)	(−1.66)	(1.17)						
1912	0	.010	−.037	−.001	.762	−3.93		.57*		
		(2.06)	(−.20)	(−.30)						
1918	6	−.013	−.729	.017	.894	−1.29			.95	
		(−2.75)	(−3.47)	(3.72)					(> .994)	
1913	0	.010	−.034	−.002	.769	−3.59				.27*
		(2.21)	(−.30)	(−.34)						(.45*)

(table continued)

(continued from the previous page)

								p−values		
T_b	k	$\hat{B}$	$\hat{\theta}$	$\hat{\gamma}$	$\hat{\alpha}$	$t_{\hat{\alpha}}$	$t^{\alpha}(t)$	$t^{\alpha}(aic)$	$t^{\gamma}(t)$	$t^{\gamma}(aic)$
(1)	(2)	(3)	(4)	(5)	(6)	(7)	(8)	(9)	(10)	(11)
				France: Model (2)						
1941	9	.004	−1.28	.014	.574	−7.75	<.01		<.01	
		(6.81)	(−7.56)	(7.59)					(<.01)	
1941	1	.003	−1.00	.011	.697	−6.66		< .01		< . − 1
		(5.72)	(−6.43)	(6.49)						(< .01)
				Germany: Model (2)						
1953	1	.005	−.155	.003	.744	−4.81	.12			
		(4.59)	(−.927)	(1.77)						
1943	1	.005	−.590	.006	.750	−4.87		.20		.03
		(4.65)	(−3.88)	(3.97)						(.06)
1943	3	.006	−.632	.008	.728	−4.63			.05	
		(4.53)	(−3.86)	(3.91)					(.09)	
				Italy: Model (2)						
1941	1	.004	−.603	.007	.767	−4.56	.21		.06	
		(4.29)	(−4.03)	(4.15)					(.11)	
1944	0	4.002	−.138	.002	.855	−3.27		.90*		
		(2.86)	(−1.01)	(1.68)						
1917	0	.002	−.136	.001	.910	−2.93				.50*
		(3.13)	(−2.67)	(1.56)						(.70*)
				Norway: Model (3)						
1941	3	.022		.020	.848	−3.18	.48			
		(57.75)		(21.80)						
1944	0	.022		.021	.876	−2.84		.68		
		(63.01)		(22.30)						
1945	0	.022		.021	.877	−2.84			.42	
		(64.68)		(22.41)					(.63)	
1945	1	.023		.021	.871	−2.85				.42
		(64.68)		(22.41)						(.63)

(table continued)

(continued from the previouspage)

T_b	k	$\hat{B}$	$\hat{\theta}$	$\hat{\gamma}$	$\hat{\alpha}$	$t_{\hat{\alpha}}$	*p−values* $t^{\alpha}(t)$	$t^{\alpha}(aic)$	$t^{\gamma}(t)$	$t^{\gamma}(aic)$
(1)	(2)	(3)	(4)	(5)	(6)	(7)	(8)	(9)	(10)	(11)
				Sweden: Model (2)						
1915	9	.006	−.086	.001	.751	−4.01	.51			
		(4.11)	(−2.72)	(2.09)						
1882	0	−0.01	−.047	.002	.899	−2.67		>.99		
		(−.27)	(−2.05)	(1.58)						
1935	9	.008	−.147	.002	.708	−3.80			.22	
		(3.91)	(−2.76)	(2.80)					(.36)	
1905	0	.003	−.033	.001	.895	−2.54				.64
				(2.35)	(−1.58)	(1.44)				(.81)
				U.K.: Model (1)						
1917	5	.006		−.086	.735	−5.59	<.01		<.01	
		(5.80)		(−5.41)					(<.01)	
1917	1	.006		−.084	.750	−.5.92		<.01		<.01
		(6.08)		(−5.42)						(<.01)
				U.S.: Model (1)						
1928	6	.022		−.074	.693	−4.58	.11		.06	
		(4.52)		(−3.04)					(.10)	
1928	1	.008	−.065	.766	−4.76		.06		.04	
		(4.69)		(−2.82)						(.06)

Notes:

The last four columns report p-values associated with the t-statistic $t_{\hat{\alpha}}$ for testing $\alpha = 1$ in the appropriate regression model denoted in the second column. $t^{\alpha}(t)$ and $t^{\alpha}(aic)$ refer to the p-values associated with choosing T_b minimizing $t_{\hat{\alpha}}$. The former chooses k using the recursive t-statistic procedure and the latter chooses k using the AIC; $t^{\gamma}(t)$ and $t^{\gamma}(aic)$ refer to the p-values associated with choosing T_b maximizing the t-statistic on the coefficient of the change in slope for Models 2 and 3 (or minimizing the t-statistic on the coefficient of the change in intercept in the case of Model 1), the former chooses k using the recursive t-statistic procedure and the latter uses the AIC. The main entries correspond to one-tailed test and those in parentheses to two-tailed tests (i.e. without imposing *a priori* the sign of the change in slope). An entry with a $*$ indicates that the residuals from the estimated equation exhibited significant serial correlation according to the Box-Pierce statistic.

The way to read the outcome of the test statistics is as follows. An entry under one of the last four columns indicates that the corresponding procedure (either $t^{\alpha}(t), t^{\alpha}(aic), t^{\gamma}(t)$, or $t^{\gamma}(aic)$) selected for the corresponding values of T_b and k listed in columns 1 and 2 of the same row.

Consider, for example, the case of Australia. Choosing T_b minimizing the t-statistic on α and choosing k using the recursive t-statistic on the last lag gave $T_b = 1943$ and $k = 2$. The associated value of the t-statistic for $\alpha = 1$ in this autoregression is -4.52 with a corresponding p-value of .03, thereby allowing a rejection of the unit root hypothesis. Using instead the *AIC* to select k gave a more parsimonious autoregression, as expected, with $k = 0$ and the date of break slightly different ($T_b = 1939$). Choosing now T_b by maximizing the t-statistic on the coefficient of the change in slope gave 1947 as the estimate of the break point using either method to select k.

However, the selected value of k is different with the two methods, the recursive t-statistic yields $k = 2$ and the *AIC* yields $k = 1$. The former methods yields $t_{\hat{\alpha}} = -4.29$ with an associated p-value of .03 if the one sided change is imposed *a priori* and of .06 if no *a priori* condition is imposed on the sign of the change. The method $t^{\gamma}(aic)$, however, yields $t_{\hat{\alpha}} = -3.44$ with associated p-values of .18 (in the one sided case) and .31 (if no *a priori* conditions are imposed on the sign). For the Australian real *GDP* series, we can therefore conclude that the unit root is rejected using either method to select T_b provided the recursive t-statistic method on the last lags is used. If the *AIC* is used to select k, one can no longer reject the unit hypothesis.

In the case of Australia, the four different procedures used implied different estimated autoregressions (i.e. values of T_b and k). This is not always the case. Consider now the real *GDP* series for Canada. If the recursive t-statistic on the last lag is used to select k both methods yield $T_b = 1929$ and $k = 4$ and the p-values associated with the corresponding unit root tests are less than .05 in both cases.[8] Using the *AIC* gave, in both cases, $T_b = 1928$ and $k = 0$ with tests having much higher p-values (.27 and .07).

---

8. For Canada, we used a slight modification. The procedures using the recursive t-statistic to select the truncation lag parameter gave $k = 7$ as the estimate (1929 still being the break date). However, the estimated autoregression in this case presented problems probably due to the presence

Note, however, that the entries for Canada corresponding to methods that use the *AIC* to select the order of the autoregression are marked with an asterisk. This denotes that the Box-Pierce statistic, as printed from the *RATS* output, indicated the presence of serial correlation in the estimated residuals at a confidence level less than .05. We therefore tend to put more weight on the results associated with $T_b = 1929$. This problem of serial correlation in the residuals does not occur in the case where the recursive t-statistic on the last lags is used to select k. It does, however, occur for several other cases when the *AIC* is used to select k. For this reason, we feel that the unit root tests based on using the *AIC* to select k may have dubious finite sample properties caused by the selection of a too parsimonious model.

Analyzing the results in general for the real *GDP* series, the following conclusions are reached. The unit root is rejected (p-values less than .10, say) for the following countries: Australia, Canada, Denmark, France, Germany, U.K. and U.S.A. (using the *AIC* yields conflicting results only for Australia and Canada). The unit root is not rejected with Model (2) for Finland, Italy, and Sweden as well as for Norway using Model (3).

With a rejection of the unit root, one can use the fact that, in the estimated autoregressions, the t-statistic on the change in slope or intercept is asymptotically normally distributed. For all these countries, this t-statistic is highly significant. Furthermore, again if the unit root is rejected, the selected value of T_b is a consistent estimate of the break point. When the unit root is not rejected $t_{\hat{\gamma}}$ is not asymptotically normally distributed and the selected values of T_b is likely to no longer yield a consistent estimate of the break point.

For the countries for which the unit root is rejected, the estimates of the break points yield interesting conclusions about important episodes which had permanent effects on the long term behaviour of the series. For France and Germany, the dates of the breaks are, as expected, associated with World War II. For

of multicolinearity (and the residuals also suggested the presence of serial correlation). We therefore decided to use the next highest value of k for which the t-statistic associated with the last lag was significant at the 5% level. Note, however, that this does not change the outcome of the unit root test. Indeed, the unit root is rejected at the 5% level for all values of k between 2 and 10.

Australia, some kind of major "take-off" occurred in the early 1940's. For Canada, the date is associated with the Great Crash of 1929, while for the U.K. it is 1917, more in line with the first World War. Finally for Denmark, the date of the break occurred in the late 1930's. Note that, for the series for which the unit root is rejected, the break date selected by the procedure $t^{\alpha}(t)$ was used to construct the broken trend function in the graphs discussed earlier.[9] When the unit root is not rejected, the selected values of T_b do not have the same properties and the choice of the brake dates used to construct the graphs in those cases was also guided by visual inspection.

4.7 CONCLUSIONS

This paper analyzed an international data set of historical real *GDP* series. We have attempted to show how allowing for a break in the trend function at a single date can alter the outcome of unit root tests for many series. The dating of these breaks are associated with major events that has a permanent effect on the behaviour of the series creating a major change in intercept (a sudden crash in the level of economic activity such as the Great Crash of 1929 for the North American countries and World War II for some European countries), a change in the slcpe of the trend function (e.g. a "take-off" for Australia in the early 1940's); or generally a combination of these events occurring around the same date.

We have discussed several testing procedures that allow treating the dating of the break as an unknown variable and referred the reader to various sources for asymptotic and finite sample critical values. Our empirical results suggest that a crucial feature of these tests is the method used to select the order of the estimated autoregressions. The outcome of the tests are sometimes dependent on which procedure is used. We tend to recommend the use of the recursive t-statistic on the last lags as opposed to an information based criteria such as the *AIC*. We are aware, however, that more work is needed to justify this preference. We hope to report analytical results on this issue in the near future.

---

9. The sole exception is Germany in which case we used the date 1943 (instead of 1953) suggested by the three other methods.

5 Wage Growth and the Inflation Process: An Empirical Approach*

Yash P. Mehra

5.1 INTRODUCTION

A popular theoretical model of the inflation process is the expectations augmented Phillips curve model. In this model, it is generally assumed that prices are set as a markup over productivity-adjusted labour costs, the latter being determined by the expected inflation rate and the degree of demand pressure.[1] It is assumed further that expected inflation depends on past inflation. This model thus implies that wages and prices are causally related with feedbacks in both directions. However, Gordon (1988) has raised some doubts on the validity of this implication. He found that changes in wages and prices are not related in the Granger-causal sense. His finding also contradicts the results in some early empirical works of Barth and Bennett (1975) in which it was shown that wages and prices are related with Granger-causality running either in both directions or only from prices to wages.

The statistical inference in the aforesaid empirical works has often been conducted under the assumption that both the wage and price series contain deterministic trends. Recent evidence has called this assumption into question and has shown that the trend components of several time series also contain stochastic elements; see Nelson and Plosser (1982) and Schwert (1988). Furthermore Stock and Watson (1989) showed that a misspecification of trend components can lead to incorrect tests of hypotheses.

---

*. This paper is an expanded and revised version of Mehra (1991).

1. This version has been closely associated with Gordon (1982, 1985, 1988). See also Stockton and Glassman (1987).

This paper, therefore, utilises some of these recent techniques to investigate the nature of trends in the wage and price series and examine if the long-run movements in prices are related to the long-run movements in the productivity-adjusted wages. It is shown that when wages and prices are measured as rates of growth they do not contain deterministic trends but share, instead, a common stochastic trend. That is, these variables are cointegrated in the sense of Engle and Granger (1987). However, the presence of the common stochastic trend, in the rates of change in wages and prices, appears to be consistent with Granger-causality, running from the rate of change in prices to the rate of change in wages, but not vice versa as suggested by the 'price markup' view. Thus, the results reported here seem to support Barth and Bennett (1975) and Mehra (1977) rather than Gordon (1988).

The plan of this paper is as follows. Section 2 presents a Phillips curve model of the inflation process and discusses its implications for the relationship between wages and prices. It also discusses how tests for cointegration and Granger-causality can be used to examine such wage-price dynamics. Section 3 presents the empirical results and Section 4 contains the concluding observations.

5.2 THE MODEL AND THE METHOD

5.2.1 A Phillips Curve Model

It has been already noted that the view that systematic movements in wages and prices are related is derived from the expectations-augmented Phillips curve model of the inflation process. To understand this, consider the following price and wage equations that typically underlie such Phillips curve models, for example, in Gordon (1982, 1985) and Stockton and Glassman (1987).

$$\Delta p_t = h_0 + h_1\Delta(w_t - q_t) + h_2 x_t + h_3 S_{pt} \tag{1}$$

$$\Delta(w_t - q_t) = k_0 + k_1\Delta p_t^e + k_2 x_t + k_3 S_{wt} \tag{2}$$

$$\Delta p_t^e = \sum_{j=1}^{n} \lambda_j \Delta p_{t-j} \tag{3}$$

where all variables are in natural logarithms, p is the price level, w_t is the nominal wage rate, q_t is labour productivity, x_t is a demand pressure variable, S_{pt}, S_{wt} are supply shocks affecting the price and wage equations respectively and Δ is the first difference operator.

Equation (1) describes the price markup behaviour. Prices are marked up over productivity-adjusted labour costs $(w - q)$ and are influenced by cyclical demand (x) and the exogenous relative price shocks (S_p). This equation implies that productivity-adjusted wages determine the price level, given demand pressures. Equation (2) is the wage equation. Wages are assumed to be a function of cyclical demand (x) and expected price level, the latter modelled as a lag on past prices as in equation (3). The wage equation implies that wages depend upon past prices, *ceteris paribus.*

The price and wage behaviour, described above, suggests that long-run movements in wages and prices must be related. Furthermore, if we allow for short-run dynamics in such price and wage behaviour, the analysis presented above would also suggest that past changes in wages and prices should contain useful information for predicting future changes in prices and wages, *ceteris paribus.* These implications could be easily examined using the tests for cointegration and Granger-causality between wage and price series.

5.2.2 Testing for Cointegration: The Engle-Granger Procedure

If the long-run components in the wage and price series are modelled as stochastic trends and if they move together, then the two time series should be cointegrated as discussed in Granger (1986). Thus, the long-run comovement of wages and prices is examined using the two-step test for cointegration proposed in Engle and Granger (1987). The first step tests whether each of the variables of interest, such as p_t and $(w_t - q_t)$, has a stochastic trend. That is investigated by performing unit root tests on the variables. The second step tests whether stochastic trends in these variables are related. That is investigated here by estimating the cointegrating regressions of the form

$$p_t = \gamma_0 + \gamma_1(w_t - q_t) + \nu_{1t} \tag{4}$$

$$(w_t - q_t) = \delta_0 + \delta_1 p_t + \nu_{2_t} \tag{5}$$

and then test whether the residual ν_1, or ν_2 above has a unit root or not. If the residual does not have a unit root, while the dependent and independent variables had each a unit root, then the price and wage series are cointegrated.[2]

5.2.3 Testing for Granger-causality: The Error- Correction Parameter

Granger (1988) also pointed out that if a pair of series is cointegrated, then there must be Granger-causation in at least one direction. This follows from the observation that such series satisfy an error-correction model of the form

$$\Delta p_t = \lambda_1 \nu_{t-1} + \text{other lags of } \Delta(w_t - q_t) \text{ and } \Delta p_t \quad (6)$$

$$\Delta(w_t - q_t) = \lambda_2 \nu_{t-1} + \text{other lags of } \Delta(w_t - q_t) \text{ and } \Delta p_t \quad (7)$$

where ν is the residual from one of the cointegrating regressions in (4) and (5) and where one of $\lambda_1, \lambda_2 \neq 0$. Since ν_{t-1} depends upon lagged levels of p_t and $(w_t - q_t)$, the above model implies that either Δp_t, or $\Delta(w_t - q_t)$, or both must be Granger-caused by the lagged levels of these variables. Hence, the presence of Granger-causality is also examined by testing whether one, or both of $\lambda_1, \lambda_2 \neq 0$. The error-correction price and wage equations (6) and (7), thus, provide an additional channel through which the relevant variables can Granger-cause prices and wages.

5.3 EMPIRICAL RESULTS

This section presents empirical results.[3] The data used are quarterly and cover the period 1959:1 to 1989:3. We have examined the dynamic interactions between wages and prices in a tri-variable system consisting of the general price level (p_t), productivity-adjusted wage $(w_t - q_t)$, hereafter denoted as $\tilde{w}_t$, and an output gap variable (g_t). The output gap variable is a proxy for demand pressures. Note that supply shocks could have important short-run effects on wages and prices. Therefore we have

---

2. The cointegrating regressions (4) and (5) are to be estimated by including only nonstationary variables. If any included explanatory variable is stationary, then the ordinary least squares procedure will give inconsistent estimates of the parameters in (4) and (5). In that case, the residual ν_{1t} or ν_{2t} is also inconsistent, thereby biasing the cointegration test results.

3. The empirical work presented here is performed using RATS (Version 3.0), VAR Econometrics, Evanston, Illinois.

conducted the Granger-causality tests by including some of these supply shocks in the tri-variable system.[4] The supply shocks considered here include the relative prices of energy, food and imports. Dummy variables for the period of the Nixon wage and price controls and for the period immediately following the wage and price controls are also included. The sample period starts from 1959 because the fixed-weight price data used here begin from that year. The definitions of the variables and the data used in this study are given in the data appendix.

5.3.1 Test Results for Unit Roots and Cointegration

The test used to investigate unit roots in price, wage, and output gap regressors is based on the augmented Dickey-Fuller test (*ADF*). It can be performed by estimating the following regression

$$\Delta y_t = \alpha + \rho y_{t-1} + \beta T + \sum_{s=1}^{n} d_s \Delta y_{t-s} + \epsilon_t \tag{8}$$

where y_t is the relevant time series and ϵ_t is the residual. T is a linear deterministic time trend and is included in (8) because the alternative hypothesis considered here is that y_t is stationary around a linear trend; see Nelson and Plosser (1982). If the series y_t has a unit root and the linear trend is not significant, then the estimated coefficients ρ and β in (8) should be zero. The results of estimating (8) for levels as well as for first-differenced regressors are presented in Table 5.1.

The t-statistics for the hypothesis $\rho = 0$ are small for $p_t, \tilde{w}_t, g_t, \Delta p_t$, and $\Delta \tilde{w}_t$, but large for Δg_t as well as for $\Delta^2 p_t$, and $\Delta^2 \tilde{w}_t$. We have not reported, however, the t-values for $\Delta^2 p_t$ and $\Delta^2 \tilde{w}_t$. These results indicate that the first differences of g_t and

4. Whether or not these supply shocks enter the wage and price regressions was first tested using the Lagrange Multiplier (*LM*) test for the omitted variables; Engle (1984). The *LM* test for k omitted variables is conducted by regressing the equation's residuals on both the original regressors and the set of omitted variables. If the omitted variables do not belong to the equation, then multiplying the R^2 statistic of this regression by the number of observations will produce a statistic asymptotically distributed as χ^2 with k degrees of freedom. These tests can be performed using RATS (Version 3).

Table 5.1

Augmented Dickey-Fuller Test Results, 1961:3-1989:3

y_t	ρ	β	Φ_3	n^1
Price Level (p_t)	−0.02(2.4)	0.14(2.4)	2.98	7
Unit Labour Cost ($\tilde{w}_t$)	−0.02(2.2)	0.33(2.2)	2.45	3
Output Gap (g_t)	−0.07(2.7)	−3.86(1.3)	3.56	2
Change in Price Level (Δp_t)	−0.13(2.3)	2.26(0.5)	2.88	6
Change in Unit Labour Cost ($\Delta\tilde{w}_t$)	−0.29(3.1)	1.17(0.2)	5.04	2
Change in Output Gap (Δg_t)	−0.60(5.3)	0.20(0.1)	13.92	1

Notes:

The absolute t-statistics of ρ and β are in the parentheses. The coefficient on trend is multiplied by 1000. The 5% critical values for the t and Φ_3 statistics are 3.45 from Table 8.5.2 in Fuller (1976) and 6.49 from Table *VI* in Dickey and Fuller (1981) respectively.

1. The parameter n was chosen by the 'final prediction error' criterion due to Akaike (1969). The Godfrey (1978) test statistics (not reported) do not indicate the presence of first, second or fourth-order serial correlation in the residuals. The absence of higher order serial correlation was also confirmed by the Ljung-Box Q statistics (also not reported).

the second differences of p_t and $\tilde{w}_t$ are stationary. Thus, p_t and $\tilde{w}_t$ series have two unit roots because these series are stationary after second differencing. In contrast, the g_t series has a single unit root because it is stationary after first differencing.

The t-statistics for the hypothesis that $\beta = 0$ are also small and imply that the price and wage series considered here do not contain time trends. The Φ_3 statistics in Table 5.1 test the joint-hypothesis that $(\rho, \beta) = (0, 0)$ and confirm the above inferences.[5]

5. The unit root tests reported in Table 5.1 were performed using lags of first-differenes of y_t, selected by the final prediction error criterion of Akaike (1969). Schwert (1988) has shown that if the variable y_t follows a mixed *ARIMA* process, then the unit root tests should be performed using somewhat more generous lag lengths. We repeated the tests with $n = 8$ and got similar results. For example, the t-values ($\rho = 0$) for variables $p_t, \tilde{w}_t, g_t, \Delta p_t, \Delta\tilde{w}_t$, and Δg_t are found to be $-2.4, -2.4, -2.3, -2.1, -2.2$ and 3.9 respectively. The t-value is significant only for Δg_t. The corresponding t-values for the trend coefficient ($\beta = 0$) are found to be

Table 5.2 presents results of testing for cointegration using levels as well as growth rates of prices and wages. Each row reports coefficients from two regressions. The first regression is the cointegrating regression of the form

$$Y_t = \alpha + \beta X_{1t} + \gamma X_{2t} + \nu_t$$

where ν_t is the residual. The second regression tests for a unit root in these residuals of the relevant cointegrating regression and is of the form

$$\Delta\nu_t = \rho\nu_{t-1} + \sum_{s=1}^{n} d_s \Delta\nu_{t-s}.$$

The coefficients reported from the first regression are α, β, and γ, and the coefficient ρ, with the absolute t-statistics in parentheses, is from the corresponding second regression.

Table 5.2

Cointegrating Regressions between Wages and Prices

Y_t	X_{1t}, X_{2t}	α	β	γ	ρ (*t-stats.*)	n^1
p_t	$\tilde{w}_t, g_t$	0.13	0.89	0.01	−0.03(1.49)	2
$\tilde{w}_t$	p_t, g_t	−0.13	1.11	−0.00	−0.03(1.65)	2
Δp_t	$\Delta\tilde{w}_t, g_t$	2.16	0.43	−0.16	−0.45(3.86)	1
$\Delta\tilde{w}_t$	$\Delta p_t, g_t$	−0.45	1.22	0.20	−0.92(7.12)	1

Notes:

Absolute t-statistics are in the parentheses. The 5% critical value for the t-statistic is 3.62; see Engle and Yoo (1987), Table 5.3.

1. The Godfrey and the Ljung-Box Q statistics, not reported, do not indicate the presence of serial correlation in the residuals of the second regression.

The estimates of the cointegrating regressions and the results of applying the *ADF* test for detecting a unit root in the residuals of these cointegrating regressions are reported in Table 5.2. These unit root tests show that while the residuals from the regressions of the levels of the variables have unit roots, those from the first difference regressions do not. These results are consistent with the inference that while levels of wage and price

2.4, 2.3, −1.3, .5, .1, and .1. None of these t-values is significant.

variables are not cointegrated, first differences are.[6] Hence, long-run movements in the growth rates of wages and prices appear to be correlated.

5.3.2 Granger-Causality Test Results From Error-Correction Models

The unit root and cointegration test results presented above imply that wage and price variables have an error-correction representation of the form

$$\Delta^2 p_t = \alpha + \sum_{s=1}^{n1} \beta_{1s}\Delta^2 p_{t-s} + \sum_{s=1}^{n2} \beta_{2s}\Delta^2 \tilde{w}_{t-s} + \sum_{s=1}^{n3} \beta_{3s}\Delta g_{t-s} + \lambda_1 \nu_{1t-1} \qquad (9)$$

$$\Delta^2 \tilde{w}_t = \gamma + \sum_{s=1}^{n1} \gamma_{1s}\Delta^2 \tilde{w}_{t-s} + \sum_{s=1}^{n2} \gamma_{2s}\Delta^2 p_{t-s} + \sum_{s=1}^{n3} \gamma_{3s}\Delta g_{t-s} + \lambda_2 \nu_{2t-1} \qquad (10)$$

where Δ^2 is the second difference operator and where ν_1 and ν_2 are the residuals from the cointegrating regressions (see Table 5.2) of the form

$$\nu_{1t} = \Delta p_t - \alpha - \beta\Delta\tilde{w}_t - \gamma g_t$$
$$\nu_{2t} = \Delta\tilde{w}_t - \delta_0 - \delta_1\Delta p_t - \delta_2 g_t.$$

Note that λ_1 and λ_2 are the error-correction coefficients.

As stated earlier if Δp and $\Delta\tilde{w}$ are cointegrated then there should be Granger-causation in at least one direction. Therefore at least one of $\lambda_1, \lambda_2 \neq 0$. Hence, even if the second differences of the price and wage regressors do not enter (9) and (10), first differences might through the residual and hence Granger-cause prices and/or wages.

Table 5.3 reports estimates of λ_1 and λ_2 (with t-statistics in parentheses) from regressions of the form (9) and (10). Since there

6. The unit root tests performed on the supply shock variables indicated the presence of a single unit root in each of the relative prices of food, energy and imports. However, including them in levels regressions of Table 5.2 does not change the result on the absence of cointegration between p_t and $\tilde{w}_t$.

Table 5.3

Error-Correction Coefficients and F Statistics for Granger-Causality

			Sample Period 1961 : 3-1989 : 3							
Lags	*From Price Regressions*					*From Wage Regressions*				
(n_1, n_2, n_3)	λ_1	$F1$	$F2$	*df*	χ^2_p	λ_2	$F3$	$F4$	*df*	χ^2_w
(8, 0, 0)	−0.05(0.6)				11.8	−0.88(5.5)				17.1
(4, 4, 4)	−0.06(0.6)	2.4[b]	4.3[a]	4, 88		−0.68(2.4)	1.5	2.3	4, 97	
(8, 4, 4)	−0.06(0.5)	2.1	3.9[a]	4, 84		−0.86(2.6)	2.3	2.3	4, 93	
(8.8.8)	−0.04(0.3)	1.6	2.7[a]	8, 76		−0.48(1.1)	1.6	2.1[b]	8, 85	
			Sample period 1961 : 3-1979 : 4							
(8, 0, 0)	0.03(0.2)				17.9	−1.02(4.7)				28.8
(4, 4, 4)	0.21(1.2)	1.1	3.6[a]	4, 49		−0.80(2.0)	1.1	1.3	4, 58	
(8, 4, 4)	0.27(1.5)	0.8	3.9[a]	4, 45		−1.02(2.1)	1.6	1.2	4, 54	
(8, 8, 8)	0.25(1.3)	0.7	2.3[b]	8, 37		−0.83(1.3)	1.0	1.2	8, 46	

Notes:

$F1$ tests all $\beta_{2s} = 0$, $F2$ all $\beta_{3s} = 0$, $F3$ all $\gamma_{2s} = 0$ and $F4$ all $\gamma_{3s} = 0$. *df* is the degree of freedom parameter for the F statistics given in the relevant row.

The price regressions also included eight past values of the relative price of energy, one past value of the relative price of imports and dummies for President Nixon's price controls.

χ^2_p is the Lagrange Multiplier (LM) test statistic for the hypothesis that 8 lags of the relative price of food and lags 2 through 8 of the relative price of imports do not enter the price regression (the 5% critical value is 25.0).

The wage regressions included only price control dummies. χ^2_w is the LM test for the hypothesis that 8 past values of relative prices of imports, food, and energy do not enter the wage regression (the 5% critical value is 36.4).

a. Significant at the 1% level.
b. Significant at the 5% level.

is some ambiguity as to how many lags of $\Delta^2 p_t$ and $\Delta^2 \tilde{w}_t$ need to be included in (9) and (10), we present results for some arbitrarily selected lag lengths. In addition, results are presented for the subperiod 1961:3 to 1979:4. The price regression (9) estimated here also included eight past values of the relative price of energy, one past value of the relative price of imports and "on" and "off" dummies for President Nixon's price controls. Other lags of the relative prices of food and imports did not enter the regression. The wage regression (10) is estimated including only the price control dummies. Relative prices of food, energy and imports were not significant in such regressions.

As can be seen from Table 5.3, λ_1 is generally not statistically significant whereas λ_2 is. Moreover, other lags of $\Delta^2\tilde{w}_t$ when included in price regressions are generally not statistically significant; see $F1$ statistics in Table 5.3.[7] These results together are consistent with the presence of Granger-causality, not from wages to prices, but from prices to wages.

It can also be seen from Table 5.3 that the output gap variable Granger-causes prices. $F2$ statistics that test the null hypothesis that lags of Δg_t do not enter price regressions are generally large, resulting in the rejection of the null hypothesis. On the other hand, $F4$ statistics that test the hypothesis that such lags do not belong in wage regressions are generally small. However, since levels of g_t do enter wage regressions through the error-correction term, which in turn is statistically significant, one cannot conclude that the output gap regressor does not Granger-cause wages. The finding that λ_2 is significant in wage regressions implies the presence of Granger-causality from lagged levels of price and output gap variables.

5.3.3 Granger-Causality Test Results from Standard Models

This section considers a specification similar in spirit to the one reported in Gordon (1988). This specification assumes that the inflation rate (Δp_t), wage growth ($\Delta\tilde{w}_t$), and output gap (g_t) variables are stationary. Hence, the presence of Granger-causality is investigated by estimating price and wage regressions of the form

$$\Delta p_t = \beta_0 + \sum_{s=1}^{n1} \beta_{1s}\Delta p_{t-s} + \sum_{s=1}^{n2} \beta_{2s}\Delta\tilde{w}_{t-s} \sum_{s=1}^{n3} \beta_{3s} g_{t-s} \qquad (11)$$

---

7. $F1$ statistic, which tests the hypothesis that $\Delta^2\tilde{w}_t$ does not Granger-cause $\Delta^2 p_t$, is significant when 4 lags of $\Delta^2\tilde{w}_t$ are included. However, this result is not robust with respect to other lag lengths or to the sub-period chosen here.

$$\Delta\tilde{w}_t = \gamma_0 + \sum_{s=1}^{n1} \gamma_{1s}\Delta\tilde{w}_{t-s} + \sum_{s=1}^{n2} \gamma_{2s}\Delta p_{t-s} + \sum_{s=1}^{n3} \gamma_{3s} g_{t-s}. \tag{12}$$

Wages do not Granger-cause prices if all $\beta_{2s} = 0$, and prices do not Granger-cause wages if all $\gamma_{2s} = 0$. These hypotheses can be tested using the standard F statistic.

Table 5.4

F Statistics for Granger-causality: The Standard Case

Lag Lengths	Price Equation			Wage Equation		
(n_1, n_2, n_3)	$F1$	$F2$	df	$F3$	$F4$	df
	Sample Period 1961 : 3-1989 : 3					
(4,4,4)	1.2	2.7[b]	4,89	5.2[a]	10.6[a]	4,98
(8,4,4)	1.2	3.8[a]	4,85	3.5[a]	11.1[a]	4,94
(8,8,8)	2.7[a]	4.0[a]	8,77	1.6	5.1[a]	8,86
	Sample Period 1961 : 3-1979 : 4					
(4,4,4)	0.2	2.3	4,50	4.3[a]	6.5[a]	4,59
(8,4,4)	0.5	3.9[a]	4,46	2.7[b]	6.2[a]	4,55
(8,8,8)	0.9	3.1[a]	8,48	1.4	3.2[a]	8,47

Notes:

The price and wage regressions included supply shock variables as noted in Table 5.3.

$F1$ tests all $\beta_{2s} = 0$, $F2$ all $\beta_{3s} = 0, F3$ all $\gamma_{2s} = 0$, and $F4$ all $\gamma_{3s} = 0$.

a. Significant at the 1% level.
b. Significant at the 5% level.

Table 5.4 presents the relevant F statistics for the price and wage regressions (11) and (12). The causality inferences using this specification are broadly similar to those inferred from Table 5.3. Wages generally do not Granger-cause prices, but prices do Granger-cause wages; see $F1$ and $F3$ statistics in Table 5.4. The output gap variable Granger-causes both prices and wages; see $F2$ and $F4$ statistics in Table 5.4.

The price and wage regressions (11) and (12) were also estimated by including a linear trend. However, the computed F statistics yielded similar inferences. That is, wages do not Granger-cause prices, but prices do Granger-cause wages. Thus,

Table 5.5

F Statistics for Granger-Causality: The Standard Case Using Trend Unit Labour Cost Variable

		Sample Period 1961 : 3-1989 : 3				
Lag Lengths	*Price Equation*			*Wage Equation*		
(n_1, n_2, n_3)	*F*1	*F*2	*df*	*F*3	*F*4	*df*
(4,4,4)	1.65	2.44[a]	4,89	6.1[a]	9.7[a]	4,98
(8,4,4)	1.44	3.35[a]	4,85	2.8[a]	11.5[a]	4,94
(8,8,8)	2.75[a]	3.83[a]	8,77	1.1	5.3[a]	8,86
		Sample Period 1961 : 3-1979 : 4				
(4,4,4)	.54	1.87	4,50	5.27[a]	6.32[a]	4,59
(8,4,4)	.55	3.13[a]	4,46	1.98	6.84[a]	4,55
(8,8,8)	.77	2.44[a]	8.38	1.21	3.79	8,47

Notes:

*F*1 tests all $\beta_{2s} = 0$, *F*2 all $\beta_{3s} = 0$, *F*3 all $\gamma_{2s} = 0$ and *F*4 all $\gamma_{3s} = 0$.

a. Significant at the 1% level.
b. Significant at the 5% level.

allowing for linear trends in (11) and (12) does not alter the main results of this paper. These results are not shown to conserve space.

Furthermore it should be pointed out that we have used in this study actual unit labour costs as a measure of productivity-adjusted wage, in contrast with Gordon who used a cyclically-adjusted trend unit labour cost variable. The latter was created by fitting a varying deterministic time trend to the labour productivity component of the unit labour cost variable.

While one can question the appropriateness of the detrending procedure chosen by Gordon, we repeated the tests of Table 5.4 using a trend unit labour cost variable similar in spirit to the one used by Gordon. Under the assumption that the labour productivity variable has a unit root, the 'trend' component of this variable was estimated using the procedure in Beveridge and Nelson (1981). We have then created a trend unit labour cost variable by deflating the wage variable (compensation per hour in non-farm business sector) by the 'trend' component of the labour productivity variable measured as output per hour in non-farm business sector. The results of performing Granger-causality tests using this unit labour cost variable are similar to the ones reported in Table 5.4. In particular, wages do not Granger-cause prices, while prices do Granger-cause wages. These results are given in Table 5. Note that the price and wage regressions used in Table 5 are of the following form respectively

$$\Delta p_t = \alpha + \sum_{s=1}^{n1} \beta_{1s} \Delta p_{t-s} + \sum_{s=1}^{n2} \beta_{2s} \Delta \hat{w}_{t-s} + \sum_{s=1}^{n3} \beta_{3s} g_{t-s}$$

$$\Delta \hat{w}_t = \gamma_0 + \sum_{s=1}^{n1} \gamma_{1s} \Delta \hat{w}_{t-s} + \sum_{s=1}^{n2} \gamma_{2s} \Delta p_{t-s} + \sum_{s=1}^{n3} \gamma_{3s} g_{t-s}$$

where $\hat{w}$ is, as explained above, the cyclically-adjusted trend unit labour cost variable.

5.4 CONCLUDING OBSERVATIONS

A central proposition in the expectations-augmented Phillips curve model of the inflation process is that prices are marked up over productivity-adjusted labour costs. If that is true, then long-run movements in prices and labour costs must be correlated. The evidence reported here shows that long-run movements in the rate of growth in prices and labour costs are correlated over time. But the presence of this correlation appears to be due to Granger-causality running from inflation to wage growth, not from wage growth to the rate of inflation. These results thus do not support the 'price markup' view of the inflation process. However, the results strongly support the implication of the original wage-type Phillips curve model in that past inflation and the past output gap help predict future wage growth. As for monetary policy, our results imply that anti-inflationary policy need not respond to labour cost data because such data contain no additional information about the future rate of inflation.

6 Diagnostic Testing: An Application to the Demand for M1

Glenn Otto

6.1 INTRODUCTION

Diagnostic or specification tests are typically used as a means of indicating model inadequacy or failure. For example in the case of a linear regression model which is estimated by ordinary least squares (*OLS*), a series of diagnostic tests could be used to indicate whether any of the assumptions required for *OLS* to be the best linear unbiased estimator (*BLUE*) appear to be violated. These assumptions include a serially uncorrelated and homoscedastic error term, absence of correlation between the error term and the regressors and correct specification of the conditional mean function, i.e. no omitted variables and appropriate functional form.

There currently exists a number of surveys on diagnostic testing in econometric models, including Pagan and Hall (1983), McAleer (1987), Beggs (1988) and MacKinnon (1992). In addition there is a textbook by Godfrey (1988) on the subject of misspecification tests. Many of these papers also discuss methodological issues in applied econometrics and the reader is referred to them for an in-depth discussion of such issues; see also Hendry and Richard (1982), McAleer, Pagan and Volker (1985) and Pagan (1987).

Applied econometric work can be viewed as consisting of a number of steps, including (i) specification of the model(s), (ii) estimation and (iii) model evaluation. Diagnostic testing plays an important role in the model evaluation stage of econometric studies. This paper is mainly concerned with some of the practical issues of diagnostic testing. An extended example, based on the demand for money, is used to illustrate some of the main aspects of diagnostic testing and also to point out some limitations and problems.

6.2 THE DEMAND FOR MONEY: AN EXAMPLE

6.2.1 Model for M1

A number of standard techniques used in diagnostic testing are illustrated using some models of the demand for money in Canada. Since it is the demand for transactions balances that is the main focus of the study, a narrow definition for money viz. $M1$ is employed. $M1$ is defined to include currency and non-interest bearing demand deposits at banks.

The demand for $M1$ in Canada has been the subject of a number of econometric studies; see Clinton (1974), Poloz (1980), Gregory and MacKinnon (1980), Gregory and McAleer (1983), Boothe and Poloz (1988) and Keil and Richardson (1990). While these studies often present the results of various diagnostic or specification tests, none provides a really systematic application of diagnostic testing.

A notable feature of many of these studies, particularly the most recent one by Keil and Richardson (1990), is the choice of sample period over which the models are estimated. Most studies only use data to the end of the 1970's. This reflects the impact of innovation in the Canadian financial system in the first half of the 1980's, which appears to have resulted in parameter instability in conventional specifications for the demand for $M1$; see Freedman (1983). Since the primary purpose of this study is to illustrate various diagnostic tests, we also estimate our models only using data to the end of the 1970's. However data from the 1980's is used in evaluating the performance of our preferred model outside of the sample of which it was estimated.

A useful survey of the main theoretical and empirical issues in the demand for money has been recently provided by Goldfeld and Sichel (1990) but much of their empirical discussion refers to work for the United States. What is also evident from Goldfeld and Sichel's survey is that the partial adjustment model (PAM) is a very common econometric specification for the demand for money. This is also evident from the above studies on Canadian $M1$. In the light of this fact we examine the adequacy of a number of $PAM's$ for Canadian $M1$.

Consider a general static (or long-run) demand for money function

$$M^* = H(P, y, R) \tag{1}$$

where $M^* =$ long-run demand for nominal $M1$, $P =$ the aggregate price level, $y =$ measure of real income (or output) and $R =$ a nominal interest rate. It is standard to assume that (long-run) demand for money functions are homogeneous of degree one, so equation 1 can be simplified to give

$$M^* = P.L(y,R) \text{ or } m = L(y,R) \tag{2}$$

where $m = (M^*/P)$. Three forms of the $L(\cdot)$ function that are commonly employed in empirical work are

$$\text{Linear}: m = \beta_1 y + \beta_2 R \tag{2.i}$$

$$\text{Loglinear}: log\ m = \beta_1\ ln\ y + \beta_2\ ln\ R \tag{2.ii}$$

$$\text{Semilog}: log\ m = \beta_1\ ln\ y + \beta_2 R \tag{2.iii}$$

where $\beta_1 > 0$, $\beta_2 < 0$ are given from economic theory; see Baumol (1952).

The log-linear specification has been the most commonly used of the three models for modeling $M1$ in the United States and Canada. However this specification constrains the long-run income and interest elasticities to be constant. There has always been some debate as to whether this is a valid assumption, particularly with respect to the interest rate. A constant interest elasticity implies that a 1% rise in the rate of interest leads to a (constant) X% fall in money demanded regardless of whether the *level* of interest rates is 5%, 10% or 20%. In contrast both the linear and the semi-log models allow the interest elasticity to vary with the level of the interest rate. For the linear model the interest elasticity is given by $\beta_2(R/M^*)$ and for the semi-log model it is given by $\beta_2 R$.

While equation 2 may provide a satisfactory explanation of long-run money holdings, in the short-run the actual stock of money balances is likely to deviate from long-run equilibrium. With quarterly data experience suggests the need to also model the short-run dynamic behaviour of $M1$. A standard dynamic specification is the partial adjustment model for nominal money balances

$$M_t - M_{t-1} = \theta(M_t^* - M_{t-1}) + v_t \tag{3}$$

where M_t^* indicates the desired long-run level on money holdings and v_t is a stochastic error term. The log and semi-log forms for $L(\cdot)$ can be accommodated by specifying equation 3 in logs.

Let us concentrate on the log-linear specification. Substituting (2.*ii*) into a log-linear version of equation 3 gives

$$ln\ m_t = \pi_1\ ln\ y_t + \pi_2\ ln\ R_t + \pi_3\ ln\ m_{t-1} + \pi_4\ ln(P_t/P_{t-1}) + v_t \quad (4)$$

where $\pi_1 = \theta\beta_1$, $\pi_2 = \theta\beta_2$, $\pi_3 = (1 - \theta)$ and $\pi_4 = -\pi_3$. Note that if $\pi_4 = 0$ then equation 4 reduces to a *PAM* for real balances; see Goldfeld and Sichel for more details. Previous empirical work for Canada suggests that a real *PAM* is consistent with the data.

The following three dynamic models are our original specification for *M*1 and they are initially estimated by (*OLS*).

$$m_t = \pi_1 y_t + \pi_2 R_t + \pi_3 m_{t-1} + \pi_4[m_{t-1} \times (\Delta P_t/P_t)] + v_{1t} \quad (5.i)$$

$$ln\ m_t = \pi_1\ ln\ y_t + \pi_2\ ln\ R_t + \pi_3\ ln\ m_{t-1} + \pi_4\ ln\ (P_t/P_{t-1}) + v_{2t} \quad (5.ii)$$

$$ln\ m_t = \pi_1\ ln\ y_t + \pi_2\ R_t + \pi_3\ ln\ m_{t-1} + \pi_4\ ln\ (P_t/P_{t-1}) + v_{3t}. \quad (5.iii)$$

6.2.2 Data and Initial Estimates

The Canadian data used to estimate models (5.*i*) to (5.*iii*) are quarterly and cover the period 1956:1 to 1978:4. All series except the interest rate are seasonally adjusted and in Table 6.1 some summary statistics are given. For ease in replicating and extending the empirical results in this paper all of the data used are reproduced in the appendix, along with their exact definitions. Plots of the data used in estimating our models for *M*1 against time (not given here), indicate that most series display persistent trends. For the purpose of this exercise it is assumed that this non-stationarity in the data can be adequately captured by a linear time trend. Thus a constant and a time trend are added to models (5.*i*) to (5.*iii*). Other sources of non-stationarity in economic data are discussed elsewhere in this book.

The coefficient estimates, their *OLS* *t*-statistics and the implied long-run income and interest elasticities are reported in Tables 6.2A, 6.2B and 6.2C. All three specifications suggest that a real *PAM* is appropriate for the current data set. This finding is at odds with results reported by Goldfeld and Sichel using quarterly data over a similar sample period for the United States. Their results indicate that for the log-linear specification the nominal *PAM* is to be preferred.

In the interest of simplicity only restricted versions of models (5.*i*) to (5.*iii*), with $\pi_4 = 0$, were estimated by *OLS*. These

Table 6.1

Means and Standard Deviations of Variables: 1956:1–1978:4

Variable	*Mean*	*Standard Deviation*	*Mean*	*Standard Deviation*
	Levels		*Logarithms*	
m : *Real M1 (\$b)*[1]	22.18	4.08	3.08	0.18
y : *Real GDP (\$b)*	202.57	65.67	5.26	0.33
R : *Interest Rate*	5.97	2.12	1.73	0.36

Notes:

1. The m and y series are in constant 1981 Canadian dollars and are seasonally adjusted at annual rates.

Table 6.2A

Linear Form of Demand for Money Function

Variable	*Coef.*	*t-stat.*	*Coef.*	*t-stat.*
y	0.007	1.68	0.007	1.79
R	−0.173	6.21	−0.177	7.24
m_{t-1}	0.879	14.64	0.868	19.73
$m_{t-1} \times (\Delta P_t / P_t)$	−0.070	−0.26	–	–
Trend	0.140	1.63	0.015	1.82
Constant	1.823	2.23	1.982	3.48
Long-run Coefficients[1]				
β_1	0.056 (0.030)[2]		0.053 (0.026)	
β_2	−1.438 (0.673)		−1.348 (0.483)	
θ	0.121 (0.060)		0.131 (0.044)	
Long-run Elasticities				
Income	0.511		0.486	
Interest	−0.387		−0.363	

Notes:
1. The long-run elasticities are calculated at the sample means.
2. The numbers in brackets are standard errors.

results are also presented in Tables 6.2A to 6.2C. Notice that the parameter estimates are little changed by the omission of the "inflation" variable from each of the models.

Table 6.2B

Log-linear Form of Demand for Money Function

Variable	*Coef.*	*t-stat.*	*Coef.*	*t-stat.*
$ln\ y$	0.196	2.49	0.199	2.54
$ln\ R$	−0.049	6.77	−0.052	7.66
$ln\ m_{t-1}$	0.825	14.38	0.790	16.68
$\Delta ln\ P_t$	−0.265	1.07	–	–
Trend	−0.001	0.73	−0.000	0.55
Constant	0.511	0.62	−0.285	0.90
Long-run Coefficients				
β_1	1.112 (0.436)[1]		0.949 (0.324)	
β_2	−0.281 (0.085)		−0.248 (0.058)	
θ	0.175 (0.057)		0.210 (0.047)	
Long-run Elasticities				
Income	1.112		0.949	
Interest	−0.281		−0.248	

Notes:

1. The numbers in brackets are standard errors.

Table 6.2C

Semi-Loglinear Form of Demand for Money Function

Variable	*Coef.*	*t-stat.*	*Coef.*	*t-stat.*
$ln\ y$	0.139	1.79	0.141	1.82
R	−0.008	6.24	−0.008	7.19
$ln\ m_{t-1}$	0.850	14.68	0.823	17.22
$\Delta ln\ P_t$	−0.219	0.85	–	–
Trend	−0.000	0.17	−0.000	0.00
Constant	−0.211	0.64	−0.143	0.45
Long-run Coefficients[1]				
β_1	0.930 (0.495)[2]		0.796 (0.384)	
β_2	−0.051 (0.019)		−0.045 (0.013)	
θ	0.150 (0.058)		0.177 (0.048)	
Long-run Elasticities				
Income	0.930		0.796	
Interest	−0.303		−0.270	

Notes:

1. The long-run elasticities are calculated at the sample means.
2. The numbers in brackets are standard errors.

Each of the models yield estimates of the key parameters $(\theta, \beta_1, \beta_2)$ which are consistent with *a priori* expectations. The estimated speed of adjustment parameter is largest for the log-linear model ($\widehat{\theta} = 0.210$) and smallest for the linear model ($\widehat{\theta} = 0.131$). These numbers point to relatively slow adjustment of actual to desired (long-run) money holdings. However this is a fairly common feature of *PAM's* which are estimated using quarterly data.

The linear and log-linear models also provide the bound for the estimated long-run income and interest rate elasticities. The estimates of the interest elasticity range from −0.36 to −0.25. A much greater variance exists for the income elasticity estimates. The linear model yields an income elasticity of about 0.5, whereas the log-linear model yields an estimated income elasticity that is not significantly different to 1.0. The semi-log model implies an income elasticity of about 0.8. For a policy maker who is interested in setting a medium term, non-inflationary growth rate for the money supply the difference between an income elasticity of 0.5 and 1.0 is of crucial importance.

All three models we have estimated yield results that are broadly consistent with economic theory. In addition the estimated coefficients of interest appear to be statistically significant. However for the *OLS* estimator to have good statistical properties a number of conditions or assumptions are required to hold. In order to check for possible deviations from these "ideal" conditions, we now perform a number of diagnostic or misspecification tests on the three models. In order to keep things simple only the results of diagnostic tests for the log-linear model are presented.

6.3 DIAGNOSTIC TESTS

The choice of diagnostic tests presented in this section is determined by what can be easily calculated using the *Microfit* econometrics computer package. This allows us to compute a wide variety of diagnostic tests. These include testing for residual serial correlation, heteroscedasticity, weak exogeneity of the regressors, functional form misspecification and parameter stability. Tests for most of these econometric problems can be directly obtained from *Microfit*. Where this is not the case and it is necessary to perform simple calculations to obtain the test statistic this is noted in the text.

6.3.1 Influential Observations

Some observations in a data set may have a relatively large effect on the results obtained in a regression model. Such data points are called influential observations since they have a disproportionally large influence on the coefficient estimates of the model. It is important to identify such observations and ensure that they are not unrepresentative of the rest of the data sample.

One measure of influence is called the *leverage* of the data. For a particular observation leverage is measured as the tth diagonal element of the $X(X'X)^{-1}X'$ matrix, where X is matrix of explanatory variables in deviations from their means, see Krasker, Kuh and Welsch (1983). A plot of the leverage of the data used in the log-linear model, which is obtained with *Microfit*, is presented in Figure 6.1. According to Krasker, Kuh and Welsch in a perfectly balanced data set each leverage point should equal (k/T), where k is the number of explanatory variables and T is the sample size. They suggest that as a rule if any leverage measure exceeds $(2k/T)$, and certainly if it exceeds $(3k/T)$, the row of data should be considered potentially influential. For our data set $(2k/T) = 0.11$ and $(3k/T) = 0.16$. Since none of the leverage measures in Figure 6.1 exceed the higher of the two bounds, there is no overwhelming evidence to suggest that any of the observations in the sample have an undue influence on the coefficient estimates. However since some of the higher leverage measures are for observations at the beginning of the sample it would be a useful exercise to examine the impact of estimating the model excluding these initial observations. Further discussion of the issue of influential observations and outliers can be found in Krasker, Kuh and Welsch and Fiebig (1987).

6.3.2 Serial Correlation

Since we are estimating our models on *time series* data it is important to check if the *OLS* residuals are serially uncorrelated. Another way of saying this is that we would like to have models that are able to produce residuals that are approximately white noise. Since all of the models contain a lagged dependent variable, so *OLS* will produce inconsistent estimates if the error terms of the models exhibit serial correlation. Even if the models

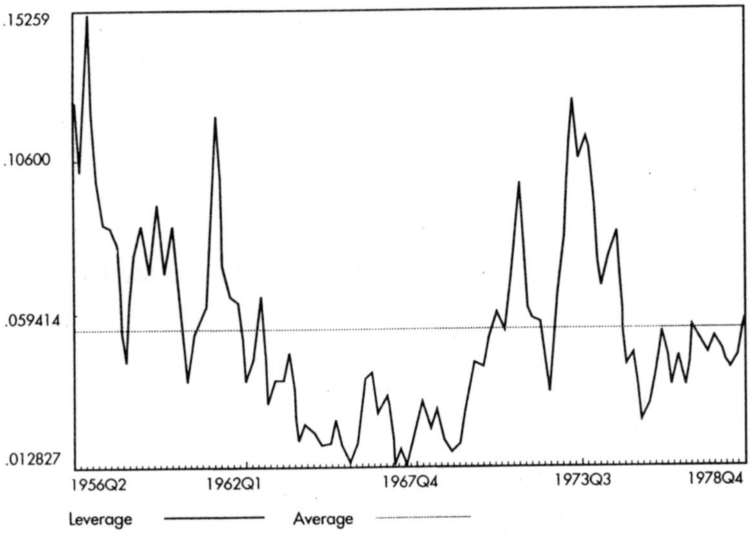

Figure 6.1 Plot of the leverage measures of the regression

were purely static, i.e. did not contain a lagged dependent variable, the *OLS* formula for the variance-covariance matrix would be incorrect in the presence of serial correlation.

Testing for serially correlated errors of order (k) in the presence of a lagged dependent variable is very straight-forward. Using the *OLS* residuals $(\hat{u}_t)$ undertake the following "artificial regression"

Regress $\hat{u}_t$ on 1 ,t ,$ln\ y_t$;$ln\ R_t$,$ln\ m_{t-1}$ and $\hat{u}_{t-1} \ldots \hat{u}_{t-k}$.

The individual t-statistics on $\hat{u}_{t-i}$ are asymptotically distributed as $N(0,1)$ and can be used to test the significance of the ith order serial correlation coefficient. A test that all coefficients on the $\hat{u}'_{t-i}s$ are jointly zero can be constructed using the F-test or as $T \times R^2$, where R^2 is the (un-centered) R^2 from the artificial regression and T is the sample size. This latter test can be motivated as a test based on the Lagrange multiplier (LM) principle; see Breusch (1978) and Godfrey (1978).

In Table 6.3 results of testing for the presence of serial correlation in the log-linear specification are reported. We test for serial correlation up to order ten. We report p-values for the joint test statistics $LM(10)$ and $F(10,76)$. The p-value indicates the probability, under the null hypothesis, of obtaining by chance a value for the test statistic that is as extreme or more extreme than the observed value of the test statistic. The smaller the p-value the stronger the evidence against the null hypothesis. The p-values for the joint tests do not indicate any evidence of serial correlation at even the 10% level of significance.

Table 6.3

Tests for Serial Correlation in the Log-Linear Model

Lag Length	*Coefficient*	*LM t-Statistic*
1	0.122	0.79
2	0.047	0.38
3	0.250	2.04*
4	−0.226	1.80
5	−0.141	1.25
6	0.164	1.34
7	−0.024	0.20
8	−0.066	0.56
9	−0.135	1.12
10	0.011	0.94
Joint Tests	$LM(10) = 14.777$ [0.140]	$F(10, 76) = 1.473$ [0.166]

Notes:

A * indicates t-statistic is significant at the 5% level of significance.
$LM(10)$ is a joint test for serial correlation up to order ten. The statistic is distributed as a $\chi^2(10)$ under the null hypothesis of no serial correlation. $F(10, 76)$ is the equivalent F-statistic version of the hypothesis test. The numbers in the [·] brackets are probability values.

Finding evidence of serial correlation in the *OLS* residuals does not necessarily imply the model errors are serially correlated. It may indicate misspecification of the model's dynamics, effects of time aggregation, parameter instability, inadequate seasonal adjustment etc. Thus it is increasingly less common for applied researchers to simply "correct" for serial correlation, by estimating a model which includes a serially correlated error term; see, however, Goldfeld and Sichel (1990) Table 8.1.

One immediate problem is that the above tests for serial correlation do not allow us to distinguish between autoregressive error (*AR*) processes and moving average (*MA*) error processes. This follows from the fact that *AR*(k) and *MA*(k) processes are *locally equivalent* under the null hypothesis of no serial correlation. Even if we take the significant t-statistics in Table 3 as indicating evidence of a serially correlated error term, we cannot say on the basis of this test alone whether it is a *MA* or an *AR* process.

However since we are in this study concerned only with the *PAM* model, there does seem to be some value in looking at the effect on the estimated parameters of modelling the error process in the log-linear model as

$$AR: \qquad u_t = \rho_3 u_{t-3} + \rho_4 u_{t-4} + \epsilon_t$$

and

$$MA: \qquad u_t = \alpha_3 \epsilon_{t-3} + \alpha_4 \epsilon_{t-4} + \epsilon_t.$$

Estimates of linear models with *AR* and *MR* error terms are easily obtained with *Microfit* and the results obtained are presented in Table 6.4. Recall that if either of the above models is true the *OLS* parameter estimates will be inconsistent and *OLS* variance-covariance matrix will be incorrect. Both of these extended models lead to more precise estimates (i.e. larger t-statistics) but do not have a major effect on the coefficient estimates. The estimated parameters for the error process are quite similar regardless of whether we use a *MA* or an *AR* process. For the model with *AR* errors there is an increase in the estimated income elasticity to around 1.50, although the magnitude of the associated standard error does not allow the rejection of a unit income elasticity.

The idea of these corrections is not to suggest these are "appropriate" models. Based on our theory we have no reason to think that they are. Rather our rational for the correction is simply to provide some indication of the robustness of the *OLS* estimates. Does the bias introduced by not accounting for the statistically significant residuals with *OLS* estimation matter from an economic viewpoint? The answer to this question would seem to be no, since correcting for serial correlation has a relatively minor effect on the results.

Table 6.4

Serial Correlation Corrections in the Log-Linear Model

	AR(3) and *AR*(4)		*MA*(3) and *MA*(4)	
Variable	*Coefficient*	*t-Statistic*	*Coefficient*	*t-Statistic*
$ln\ y$	0.316	3.53	0.228	7.88
$ln\ R$	−0.053	7.80	−0.052	10.13
$ln\ m_{t-1}$	0.788	15.26	0.788	16.22
Trend	−0.002	1.97	−0.001	5.65
Constant	−1.543	1.70	−0.524	2.89
$\hat{u}^1_{t-3}$	0.243	2.27	0.184	1.89
$\hat{u}_{t-4}$	−0.235	2.04	−0.224	2.01
Long-run Coefficients				
β_1	1.490 (0.402)[2]		1.075 (0.115)	
β_2	−0.251 (0.061)		−0.245 (0.055)	
θ	0.212 (0.052)		0.212 (0.048)	

Notes:

1. Estimated *OLS* residuals lagged as indicated.
2. Values in brackets are standard errors.

6.3.3 Heteroscedasticity

While heteroscedasticity does not in the present context lead to inconsistency of *OLS*, it does mean that *OLS* coefficient estimates are relatively inefficient. More importantly the *OLS* formula for the variance-covariance matrix is no longer valid. So inference cannot proceed using *OLS* standard errors. A standard response to this problem is to use an estimate of the variance-covariance matrix that is consistent in the presence of heteroscedasticity of unknown form. Such an estimator was developed by White (1980). In Table 6.5 the *OLS* and White's heteroscedasticity consistent t-statistics are presented for the log-linear model. It can be seen from the last column that the differences between the two sets of t-statistics are minimal, which suggests that heteroscedasticity does not pose a major problem for inference in this case. Of course this is a relatively informal approach to detecting the presence of heteroscedasticity.

Table 6.5

Comparison of OLS and Heteroscedasticity Robust t-Statistics: Log-linear Model

Variable	*OLS* *t-Statistic*	*White's* *t-Statistic*	*Ratio*[1]
$ln\ y$	2.535	2.809	1.11
$ln\ R$	7.663	7.663	1.00
$ln\ m_{t-1}$	16.684	15.701	0.94
Trend	0.548	0.640	1.17
Constant	0.271	0.323	1.19

Notes:
1. Ratio of *OLS* *t*-statistic to White's *t*-statistic.

Formal tests for heteroscedasticity based on the *LM* principle have been developed by Godfrey (1978) and Breusch and Pagan (1979) under the assumption of normal error terms. Koenker (1981) developed a test procedure which is valid even when the error terms are non-normal. The alternative hypothesis for such tests is that error term in (5.*ii*) has the property,

$$E(u_t^2) = h(\mu + Z_t\gamma)$$

where Z_t is a $(1 \times m)$ vector of functions of the regressors. Under the null $\gamma = 0$ and u_t has a constant variance.

The test is based on the following "artificial" regression;

$$\text{Regress } \hat{u}_t^2 \text{ on } 1\ , Z_{t1} \ldots Z_{tm}$$

where $(T \times R^2)$ from this regression will be asymptotically distributed as $\chi^2\ (m)$ under the null that $\gamma = 0$.

The dependent variable $\hat{u}_t^2$ is the square of the *OLS* residuals from the log-linear model. However the choice of Z_{ti} is somewhat arbitrary. Inn our case some reasonable candidates for Z_{ti} include: the predicted values and their squares i.e. $ln\ \hat{m}_t$ and $(ln\ \hat{m}_t)^2$, the regressors and their squares or some sub-set of these variables. While these tests for heteroscedasticity are not directly computed by *Microfit* they can be easily obtained via a few simple steps. We report the results obtained from the *LM* and *F*-tests in Table 6.6. The *p*-value for the test based on the regressors and their squares indicates that both the *LM* and the F statistics are significant at the 5% level, providing evidence against the null hypothesis of constant error variance. If we examine some of the regressors and their squares individually, the results, reported in rows (iii) and (iv), suggest the error variance may be related to the interest rate. Although not pursued here it may be interesting to re-estimate the log-linear model by generalized least squares (*GLS*), with a correction for heteroscedasticity implied by row (*v*) of Table 6.6.

Table 6.6

Lagrange Multiplier Tests for Heteroscedasticity, ARCH and Normality: Log-Linear Model

	Variable	*LM Statistic*	*F-Statistic*
(*i*)	*Predicted values and their squares*	$\chi^2(2) = 2.47$ [.29]	$F(2,88) = 1.23$ [.30]
(*ii*)	*Regressors and their squares*	$\chi^2(8) = 15.2$ [.05]	$F(8,89) = 2.06$ [.05]
(*iii*)	*Trend*, $(Trend)^2$	$\chi^2(2) = 2.45$ [.29]	$F(2,88) = 1.21$ [.30]
(*iv*)	$ln\ y$, $(ln\ y)^2$	$\chi^2(2) = 2.48$ [.24]	$F(2,88) = 1.42$ [.25]
(*v*)	$ln\ R, (ln\ R)^2$	$\chi^2(2) = 12.3$ [.00]	$F(2,88) = 6.85$ [.00]
(*vi*)	*ARCH* (10)	$\chi^2(10) = 11.7$ [.30]	$F(10,76) = 1.12$ [.36]
(*vii*)	*Normality*	$\chi^2(2) = 2.92$ [.23]	–

A form of heteroscedasticity that is frequently found in high frequency financial data is autoregressive conditional heteroscedasticity (*ARCH*). In this case the Z_t vector is composed of $\hat{u}_{t-1}^2 \ldots \hat{u}_{t-k}^2$. Table 6.6 contains the results of an *LM* test for *ARCH* up to order 10. The test statistics indicate that *ARCH* is not a problem in this model. Also reported in Table 6.6 is a test for normality of errors due to Jarque and Bera (1981). This is a joint test for the absence of skewness and for kurtosis equal to three. The computed test statistic indicates that it is not possible to reject the assumption of normal errors. Both the test for *ARCH* errors and normality of errors are computed directly by *Microfit*. As with other forms of heteroscadasticity neglecting *ARCH* when it is present in the model will invalidate the usual *OLS* variance-covariance matrix. While many of the estimators and tests used in this study have only a large sample (or asymptotic) justification and do not require the error term to be normally distributed, the Jarque and Bera test is useful fir indicating cases when the distribution of the error term is distinctly non-normal. In these circumstances a relatively large sample may be required for finite sample distributions to closely approximate their asymptotic equivalents.

6.3.4 Weak Exogeneity

Consistency of the *OLS* estimator requires that the regressors in the model are (asymptotically) uncorrelated with error term. More specifically in the case of the log-linear model

$$ln\ m_t = \pi_1\ ln\ y_t + \pi_2\ ln\ R_t + \pi_3\ ln\ m_{t-1} + v_t \tag{6}$$

we require

$$plim\ [T^{-1} \sum_{t=1}^{T} X_t' v_t] = 0$$

where $X_t' = (ln\ y_t, ln\ R_t, ln\ m_{t-1})$. Provided v_t is serially uncorrelated the above condition will be satisfied for $(ln\ m_{t-1})$ since it is a pre-determined variable. However it is not unreasonable to suspect that both real income and the nominal interest rate might be endogenous variables, and therefore correlated with the error term.

In the current context we wish to know if the consequences of such endogeniety (or simultaneous equation bias) has a significant impact on the *OLS* estimator. Tests for exogeneity based on a comparison of the *OLS* estimates with those obtained from an instrument variable estimator (*IV*), were initially suggested by Durbin (1954) and further developed by Wu (1973), Hausman (1978) and Nakamura and Nakamura (1981).

In the context of the linear regression model testing for exogeneity simply requires a variable addition test. Treating $ln\ y_t\ ln\ R_t$ as potentially endogenous we regress each on a set of instruments, save the residuals from the two regressions and add them to the log-linear model 6. The joint significance of the additional regressors can be tested via an F-test. Under the null of exogeneity the extra regressors should be insignificant.

Performing the above test using (a constant, linear trend, $ln\ y_{t-1}$, $ln\ R_{t-1}$ and $ln\ m_{t-1}$) as instruments yields an F-statistic of $F(2,84) = 0.412$, which has a p-value of 0.663. This suggests that possible endogeniety of real income and the interest rate does not lead to inconsistent *OLS* estimates. Note that the R^2s obtained from regressing $ln\ y_t$ and $ln\ R_t$ on the instruments are 0.99 and 0.89 respectively.

As a final check Table 6.7 presents generalized instrumental variables (*GIV*) estimates of the log-linear model. In this case the instruments used were (a constant, linear trend, $ln\ y_{t-1}$, $ln\ y_{t-2}$, $ln\ R_{t-1}$ $ln\ R_{t-2}$, $ln\ m_{t-1}$ and $ln\ m_{t-2}$). The *GIV* coefficient estimates are very similar to those obtained by *OLS*, (see Table 6.2B). The *GIV* estimator imposes three over-identifying restrictions, the validity of which can be tested via Sagan's χ^2 test. The test statistic is a $\chi^2(3) = 1.056$ has a p-value of 0.788, which points to the validity of the instruments chosen.

Table 6.7

GIV Estimates of Log-Linear Demand for Money

Variable	*Coefficient*	*t-Statistics*
$ln\ y$	0.240	2.27
$ln\ R$	−0.053	6.05
$ln\ m_{t-1}$	0.782	15.25
Trend	−0.001	0.82
Constant	−0.452	1.07
Long-run Coefficients		
β_1	1.102 (0.298)[1]	
β_2	−0.245 (0.058)	
θ	0.218 (0.051)	
Sargan's Over-identification Test :	$\chi^2(3) = 1.056$ [0.788]	

Notes:
1. The numbers in brackets are standard errors.

6.3.5 Functional Form

Tests for functional form tend to fall into two categories. One involves testing the model at hand against some *non-specific* alternative model. The second approach is to test the model against a fully specified *non-nested* alternative model. The basic idea is that if the model under consideration is correctly specified (i.e. a good approximation to the data generating process) then adding extra variables to the *conditional mean function* should not significantly improve the performance of the model. The power of these tests will depend upon choice of the variables to be added to the model.

(i) Non-specific alternatives

The most commonly used test for misspecification of the conditional mean function is Ramsey's (1969) *RESET* test. This test involves augmenting the regression model with powers of the predicted values from the model and is computed by *Microfit*. The significance of these additional variables is then tested using a standard F-test. Under the null hypothesis of no mispecification these additional variables should be insignificant, whereas if the additional regressors are statistically significant this might be viewed as evidence of model misspecification. Table 6.8 presents the results of *RESET* tests which employ the second and third powers of the predictions from the log-linear model. Neither version of the test points to any evidence of misspecification at conventional significance levels.

Table 6.8

Non-Specific Tests for Functional Form
Log-Linear Model

Variables	*LM Statistic*	*F-Statistic*
RESET		
$(ln\ \hat{y})^2$	$\chi^2(1) = 0.05$ [.83]	$F(1,85) = 0.05$ [.83]
$(ln\ \hat{y})^2, (ln\ \hat{y})^3$	$\chi^2(2) = 2.98$ [.22]	$F(2,84) = 1.43$ [.25]
Differencing Test	$\chi^2(3) = 0.37$ [.95]	$F(3,81) = 0.11$ [.93]

Another test for misspecification which is easily performed, although it is not directly calculated in *Microfit*, is the *differencing* test due to Plosser, Schwert and White (1982). If a linear (in parameters) regression model is correctly specified then asymptotically the same parameter estimates should be obtained whether the model is specified in *levels* or in *first differences*.

For example if the log-linear model is correctly specified then it should not matter whether we estimate;

$$ln\ m_t = \pi_1\ ln\ y_t + \pi_2\ ln\ R_t + \pi_3\ ln\ m_{t-1} + v_t \qquad (7.i)$$

or

$$\Delta ln\ m_t = \pi_1\ \Delta ln\ y_t + \pi_2\ \Delta ln\ R_t + \pi_3\ \Delta ln\ m_{t-1} + \Delta v_t \qquad (7.ii)$$

where Δ is the first difference operator. Notice, however, that the differenced version of the *PAM* cannot be consistently estimated by *OLS*. If (7.*i*) is correct then the error term in (7.*ii*) is serially correlated (i.e. a non-invertible $MA(1)$ process) and in conjunction with the lagged dependent variable *OLS* will yield inconsistent parameter estimates.

One solution to this problem is to use an instrumental variables (*IV*) estimator for (7.*ii*), using $\Delta\ ln\ m_{t-2}$ as an instrument. However Breusch and Godfrey (1986) have shown that the differencing test can be computed as a variable addition test on the levels model. The additional variables are constructed from the existing regressors by adding the observation lagged one period to the observation lead forward one period. Thus we would augment (7.*i*) with $[(ln\ y_{t+1} + ln\ y_{t-1}), (ln\ R_{t+1} + ln\ R_{t-1})$ and $ln\ m_{t-2}]$. Notice that in the case of the lagged dependent variable we include $ln\ m_{t-2}$ and not $(ln\ m_t + ln\ m_{t-2})$. Two sample observations are lost in constructing the regressors. The specification test simply amounts to an F or a χ^2 test for the joint significance of the additional regressors. The results obtained from applying this test to the log-linear money demand equation are reported in Table 6.8. As with the *RESET* test there is no evidence of model misspecification.

(ii) Specific non-nested alternatives

Two regression models are said to be *non-nested* (or separate) when neither model can be obtained from the other under appropriate restrictions. The three models of money demand considered in this paper are examples of non-nested models. All three models are however special cases of a more general model called the Box-Cox transformation; see Box and Cox (1964). Godfrey and Wickens (1981) suggest a means of testing the linear and log-linear specifications against the general Box-Cox model using the Lagrange Multiplier principle. Since we are only concerned with the special cases of the more general model, there seems little to be gained from testing against unrestricted Box-Cox transformation.

The concept of non-nested testing is due to Cox (1961, 1962) and was adapted for regression models by Pesaran (1974) and Pesaran and Deaton (1978). Since then a large number of non-nested tests have been developed; see the survey by McAleer (1986). The purpose of a non-nested testing procedure is to see if any of the models under consideration provide evidence which indicates that one or all of the other models are misspecified. Instead of simply testing against the data, the maintained model is tested against an alternative model.

A commonly used non-nested test is the J-test of Davidson and MacKinnon (1981). This test is appropriate when the models under consideration have the same dependent variable, in our case the log-linear and the semi-log specifications. If we initially choose the log-linear model as the null hypothesis, i.e. it is the model we are assuming to be correctly specified, then we have

$$H_0: \quad ln\ m_t = \pi_1\ ln\ y_t + \pi_2\ ln\ R_t + \pi_3\ ln\ m_{t-1} + v_{0t}$$

$$H_1: \quad ln\ m_t = \pi_1\ ln\ y_t + \pi_2 R_t + \pi_3\ ln\ m_{t-1} + v_{1t}.$$

To implement the J-test estimate the alternative model H_1 by OLS and add the predicted valued $(ln\ \hat{m}_t)$ to null model H_0. This augmented model is then estimated by OLS and the t-statistic on $(ln\ \hat{m}_t)$ is asymptotically distributed as $N(0,1)$ under the null hypothesis H_0.

Although not computed in the same manner, other non-nested procedures used here to test the log-linear and the semi-log specifications are the N test of Pesaran and Deaton (1978), the JA-test of Fisher and McAleer (1981) a test based on a *double-length regression* (DL) due to Davidson and MacKinnon (1985). All of these tests also are distributed as $N(0,1)$ in large samples.

To test linear model against the log-linear and semi-log models somewhat different tests are available. One such test is the PE test developed by MacKinnon, White and Davidson (1983). Suppose we wish to test log-linear against the linear model, i.e.

$$H_0: ln\ m_t = \pi_1\ ln\ y_t + \pi_2\ ln\ R_t + \pi_3\ ln\ m_{t-1} + v_{0t}$$

$$H_1: m_t = \pi_1 y_t + \pi_2 R_t + \pi_3 m_{t-1} + v_{1t}.$$

Let $(ln\ \hat{m}_t)$ and $(\tilde{m}_t)$ be the predicted values of the dependent variables under H_0 and H_1 respectively. The *PE* test is based on the following test regression;

$$\text{Regress } \hat{v}_{0t} \text{ on } ln\ y_t,\ ln\ R_t,\ ln\ m_{t-1} \text{ and } [\tilde{m}_t - \exp(ln\ \hat{m}_t)]$$

where $\hat{v}_{0t} = (ln\ m_t - ln\ \hat{m}_t)$ and the significance of the regressor $[\tilde{m}_t - \exp(ln\ \hat{m}_t)]$ is tested via a standard t-test. To test the linear model against the log-linear specification the following regression is run

$$\text{Regress } \tilde{v}_{1t} \text{ on } y_t, R_t, m_{t-1} \text{ and } [\hat{m}_t - ln\ \tilde{m}_t]$$

and the significance of $[ln\ \hat{m}_t - ln\ \tilde{m}_t]$ tested. Some alternative tests for linear *vs.* log-linear models that have been developed include the *BM* test of Bera and McAleer (1989) and Davidson and MacKinnon's *DL* based test.

Table 6.9 provides the results of a number of paired non-nested tests in which each of the models was tested against the other. Most of the tests indicate that the semi-log model can be rejected against the log-linear model. The exception is the *DL* test which indicates that both models are rejected by each other. All of these tests are available in *Microfit*.

Table 6.9

Non-Nested Tests of Money Demand Functions

Tested Model	Test Statistics[1]			
	N	*J*	*JA*	*DL*
Log-linear	−0.22	0.22	0.22	2.27
Semi-log	−2.33	2.10	2.10	3.12
	PE	*BM*	*DL*	
Log-linear	0.39	−0.40	2.37	
Linear	2.16	0.95	4.31	
Semi-log	1.61	1.31	2.29	
Linear	1.14	−0.13	3.68	

Notes:

1. All test statistics are asymptotically distributed as $N(0,1)$ under the null hypothesis.

Testing the log-linear and the linear specifications against each other yields results which depend on the test. The *PE* test suggests the log-linear model is to be preferred. According to the *BM* test neither model provides any evidence against the other, while the *DL* again points to the rejection of both models. Finally for the linear verses the semi-log models the *PE* and *BM* tests lead to the rejection of neither model at the 5% level, while the *DL* test again rejects both models.

In summary, aside from the *DL* test, the results of the non-nested tests point to the log-linear model as the preferred specification for Canadian money demand.

6.3.6 Parameter Stability Tests

A common means of testing for parameter stability in the absence of prior information about the exact timing a structural break is to use tests developed by Brown, Durbin and Evans (1975). These tests are based on the cumulative sum of the recursive residuals (*CUSUM*) and the cumulative sum of recursive residuals squared (*CUSUMSQ*). Figure 6.2 presents the results of *CUSUMSQ* test for the log-linear model along with the 95% confidence interval. Such bounds are not strictly valid for models with lagged-dependent variables but the test may still provide some informal evidence about parameter stability. In fact it provides little or no obvious evidence of parameter instability or misspecification in the log-linear model.

A formal test for parameter stability which is valid for models with lagged dependent variables is called the *predictive failure* test; see Chow (1960). This is sometimes called Chow's second test. The model under consideration is estimated on a sub-sample of the available data set and these coefficient estimates then used to obtain predictions over the remainder of the sample, say (p) observations. The sum of the p squared errors in prediction, divided by a consistent estimate of the error variance from the sub-sample, is (asymptotically) distributed as χ^2 with p degrees of freedom.

Since we have only used a data up to 1978:4 to estimate the log-linear model, but have data available to 1988:4, we have a relatively large number of observations with which to compute the stability test. For $p = 20$ we obtain $\chi^2(20) = 40.246$ and for $p = 40$ we get $\chi^2(4) = 70.618$, which have p-values of 0.005 and 0.002 respectively. These tests provide strong evidence against the hypothesis of model adequacy *outside* of the initial estimation period.

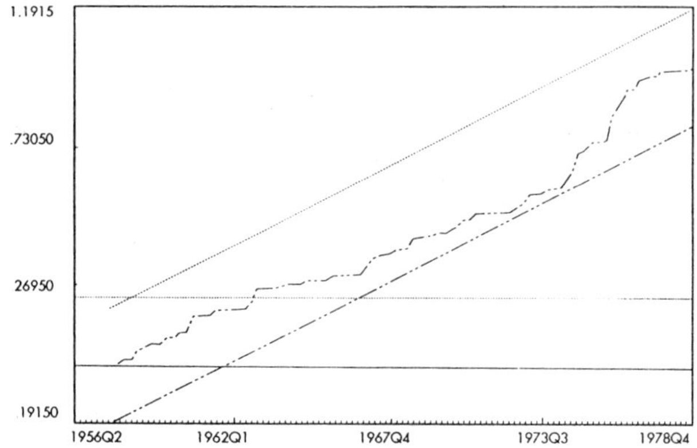

Figure 6.2 Plot of cumulative sum of squares of recursive residuals

In this particular case the failure of all three partial adjustment models, once data for the 1980's are used, is readily apparent from looking at the dynamic forecasts of the models. Figures 6.3 and 6.4 present the dynamic forecasts of the log-linear and the semi-log models for the period 1979:1 to 1988:4. While both models end up over predicting real $M1$ demand by a large magnitude, the semi-log specification provides considerably accurate forecasts for about twelve quarters after 1978:4.

This difference in performance of the two specifications was noted by Freedman (1983,p.105). He argued that "with respect to functional form, ..., until recently there has been very little difference between the semilogarithmic and logarithmic equations ... The crucial experiment which would have allowed one to discriminate between the two equations was the high interest rate environment of the last three years." However according to Freedman this was exactly when these demand for money functions where subject to structural shifts. A claim which is confirmed for the log-linear model by the above tests.

In addition according to Freedman; "If one uses a semilogarithmic equation, much of the slow growth in $M1$ ... can be attributed to the level of interest rates with less room for a shift. ... (W)ith a double-logarithmic specification one explains much less of the slow growth of $M1$ by the level of interest rates and much more via the shift." The dynamic forecasts in Figures 6.3 and 6.4 clearly support Freedman's argument, however with the

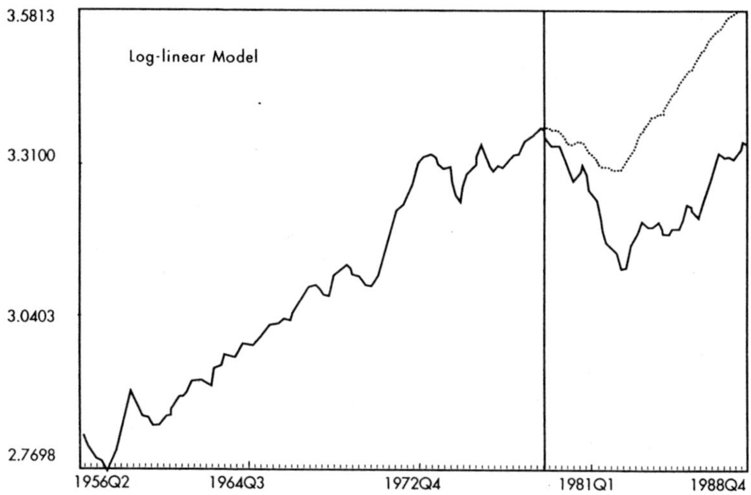

Figure 6.3 Plot of actual and dynamic forecasts
LMR ——— *Forecast* ....

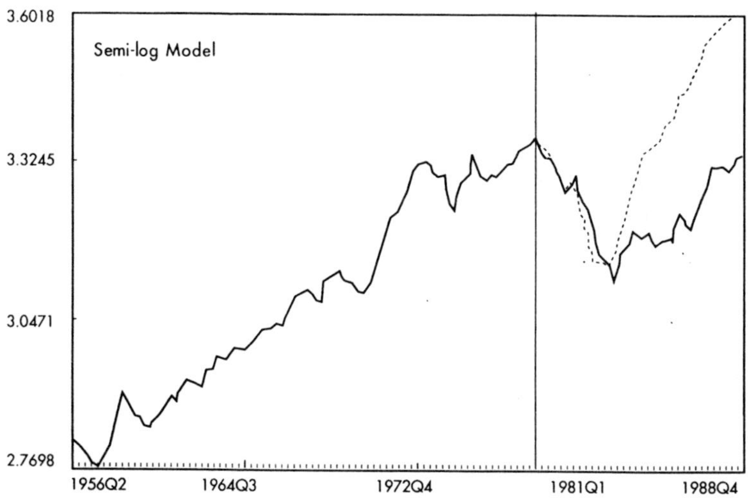

Figure 6.4 Plot of actual and dynamic forecasts
LMR ——— *Forecast* ....

availability of more data for the 1980's it is clear that even the semi-log model indicates evidence of a structural shift in the demand for $M1$.

6.4 CONCLUSION

In the preceding section we have examined the ability of a log-linear model of money demand to withstand a battery of diagnostic and specification tests. The performance of the log-linear model over the sample period 1956:1 to 1978:4 would seem to be quite reasonable, in that most of the diagnostic tests employed do not provide any strong evidence of model inadequacy. However once data for the 1980's is introduced, both the predicative failure test and the dynamic forecasts indicate that the log-linear specification does not provide an adequate model for Canadian $M1$. This result would seem to be a good illustration of the usefulness of testing the performance of any model against new data.

Data Appendix

In this appendix data used in all the papers of this volume are tabulated. Definitions of the variables are given at the end of each table. Some of these tables are long and run into several pages.

An electronic version of these data can be obtained through the usual international computer network systems. Readers should first consult the computer services depart of their own institutions for information on how to access the University of New South Wales computer network system. The data files are available from "DEMON.CSD.UNSW.EDU.AU" via "ANONYMOUS.FTP". The following commands should be issued after logging into the University of New South Wales network:

```
FTP DEMON.CSD.UNSW.EDU.AU
login: FTP
Password: [e-mail address]
FTP> CD PUB
FTP> CD RAO
FTP> ASCII
FTP> get MONEY.DAT (file name must be in upper case)
FTP> bye (to log-off)
```

In Tables D.1 to D.2 quarterly time-series data on 14 macro variables, used by Dickey, Jansen and Thornton in Chapter 2, are tabulated.

In Table D.3 quarterly U.K. data used by Holden and Perman in Chapter 3 to estimate their consumption function are given. Three time-series variable and three dummy variables are given in these tables. The following additional notes explains how these data have been collected.

Consumption, income and wealth series in Table D.3 are derived from the following original series:

(i) *CAAB*–seasonally adjusted real consumers' expenditure in millions of pounds, 1985 prices.

(ii) *AIIX*–seasonally adjusted nominal consumers' expenditure in millions of pounds, current prices.

(iii) *AIIW*–seasonally adjusted nominal personal disposable income in millions of pounds, current prices. The three series can be found in Table 5 of *Economic Trends Annual Supplement* (1992).

The wealth series is defined as seasonally adjusted gross personal financial wealth in millions of pounds, current prices, and is compiled from several editions of *Financial Statistics* in order to obtain a consistent series.

Constant (1985) price series for income and wealth were obtained from the current price series using the implicit consumers' expenditure deflator ($AIIX/CAAB$). C, I and W are the natural logarithms of the constant price consumption, income and wealth series.

The dummy variables are defined as follows:
$DD682 = 1$ in 1968 : 2, -1 in 1968 : 3 and 0 otherwise.
$DD792 = 1$ in 1979 : 2, -1 in 1979 : 3 and 0 otherwise.
$DD883 = 0$ up to and including 1988 : 2 and 1 thereafter.

In Chapter 4 Perron has used historical time-series on annual aggregate output for 11 countries. These data are given in Tables D.4 and D.5.

In Chapter 5 Mehra has estimated quarterly wage-price equations for the U.S.A. Data on 8 variables are tabulated in Table D.6. The following additional notes explain further details of these data:

RGNP is real *GNP*; *PGNP* potential real *GNP*; *ULC* unit labour cost; *GDF* the fixed-weight *GNP* deflator; *GDFIM* the fixed-weight import deflator; *GDFCF* the fixed-weight deflator for food in personal consumption expenditure; *GDFCE* the fixed-weight deflator for energy in personal consumption expenditure; and *GDFCO* the fixed weight deflator for personal consumption expenditure excluding food and energy.

All the data are from Citibank data base with the exception of *PGNP*, which is from the Board of Governors of the Federal Reserve System. This series is given in Jeffrey J. Hallman *et. al* (1990).

The variables used in the study are defined as follows where ln below is the natural logarithm of the variable.

$g_t = \ln(RGNP/PGNP)_t * 100.0$

$\tilde{w}_t = ln(ULC)_t$

$p_t = \ln(GDF)_t$

Relative Price of Energy $= \ln(GDFCE/GDFCO)_t$

Relative Price of Food $= \ln(GDFCF/GDFCO)_t$

Relative Import Prices $= \ln(GDFIM/GDF)_t$

President Nixon's Price and Wage Controls:

$PC1 = 1$ from 71(3) *to* 72(4); 0 otherwise.

$PC2 = 1$ from 73(1) *to* 74(4); 0 otherwise.

The 'trend' component of the labour productivity variable was defined as the current observed value of the variable plus all forecastable future changes in this variable beyond the mean rate of drift; see Beveridge and Nelson (1981, p.9). In particular, the 'trend' component is computed as in (a) below.

$$\tilde{q}_t = q_t + \lim_{k \to \infty} \{[\Delta\hat{q}_t(1) + \Delta\hat{q}_t(2) + \cdots + \Delta\hat{q}_t(k)] - ku\} \qquad (a)$$

where q_t is actual labour productivity; $\tilde{q}$ the trend component; $\Delta\hat{q}_t(k)$ is k-period ahead forecast of change in labour productivity; and u the mean rate of drift. The future k-period forecasts were generated under the assumption that Δq_t follows an autoregressive model of the form (b).

$$\Delta q_t = u + \sum_{s=1}^{12} b_s \Delta q_{t-s} \qquad (b)$$

where u is the constant term. The parameter k in equation (b) was set at 40. Thus, each period the equation (a) was estimated, forecasts prepared and summed as in (a) so as to generate an observation for $\tilde{q}_t$. The resulting time series $\tilde{q}_t$ is then used, as indicated above, to create the trend unit labour cost variable.

Finally Otto has used the Canadian quarterly data to estimate his demand for money function and his data on 4 variables are given in Table D.7.

Table D.1

Data Used in Chapter 2: 1953:1 to 1988:4

Period	*K*	*KSA*	*R3M*	*R10Y*	*RGNP*
1953:1	-1.22686	-1.22683	0.6831	*NA*	7.26641
1953:2	-1.20181	-1.20216	0.76686	1.09761	7.2758
1953:3	-1.22454	-1.22275	0.67141	1.07056	7.27115
1953:4	-1.21438	-1.21707	0.3873	0.97191	7.26305
1954:1	-1.22605	-1.22582	0.05827	0.892	7.24907
1954:2	-1.22598	-1.22504	-0.23953	0.85314	7.24508
1954:3	-1.24536	-1.24511	-0.12443	0.85314	7.257
1954:4	-1.27515	-1.27537	0.01686	0.90543	7.27156
1955:1	-1.27538	-1.276	0.20457	0.97191	7.29275
1955:2	-1.27897	-1.27775	0.39407	1.01632	7.30364
1955:3	-1.27372	-1.27315	0.61896	1.08079	7.31688
1955:4	-1.26812	-1.2685	0.85015	1.06815	7.32561
1956:1	-1.27093	-1.27077	0.84458	1.06471	7.32363
1956:2	-1.27519	-1.27363	0.94274	1.1259	7.32817
1956:3	-1.26832	-1.26777	0.94895	1.18571	7.3289
1956:4	-1.26726	-1.26633	1.10955	1.24502	7.33993
1957:1	-1.26235	-1.26255	1.1314	1.22466	7.34814
1957:2	-1.26359	-1.26177	1.14327	1.28841	7.34756
1957:3	-1.25523	-1.25589	1.20985	1.36788	7.3534
1957:4	-1.24984	-1.24998	1.19483	1.29006	7.33778
1958:1	-1.25641	-1.25691	0.56531	1.11186	7.31728
1958:2	-1.27845	-1.27556	-0.04395	1.07261	7.32264
1958:3	-1.28127	-1.28293	0.51879	1.25276	7.34601
1958:4	-1.28207	-1.282	0.98954	1.335	7.36941
1959:1	-1.35216	-1.29197	1.01993	1.38379	7.38175
1959:2	-1.35508	-1.29017	1.09861	1.44856	7.40062
1959:3	-1.35877	-1.28696	1.26413	1.50474	7.39603
1959:4	-1.35036	-1.28608	1.4422	1.52235	7.40452
1960:1	-1.34675	-1.27424	1.35403	1.50118	7.42154
1960:2	-1.34434	-1.27284	1.09628	1.44927	7.41866
1960:3	-1.35867	-1.28797	0.85866	1.34365	7.41962
1960:4	-1.35659	-1.28659	0.83595	1.35764	7.41101
1961:1	-1.36546	-1.29158	0.85441	1.33157	7.42136
1961:2	-1.37491	-1.29116	0.83421	1.33237	7.43373
1961:3	-1.37352	-1.29713	0.83421	1.38128	7.44793
1961:4	-1.37163	-1.30664	0.90016	1.37952	7.47017

Period	*K*	*KSA*	*R3M*	*R10Y*	*RGNP*
1962:1	-1.3669	-1.28871	1.00173	1.39053	7.48319
1962:2	-1.3631	-1.28762	0.99806	1.35403	7.49354
1962:3	-1.35407	-1.28679	1.0438	1.38379	7.50279
1962:4	-1.35148	-1.2826	1.03425	1.36174	7.50114
1963:1	-1.34935	-1.35115	1.06712	1.35918	7.51458
1963:2	-1.34073	-1.3411	1.07739	1.377	7.52833
1963:3	-1.33479	-1.33404	1.1918	1.39451	7.54565
1963:4	-1.32813	-1.32997	1.2519	1.41585	7.55281
1964:1	-1.31965	-1.32183	1.2613	1.43031	7.57492
1964:2	-1.30846	-1.30794	1.24617	1.43508	7.58345
1964:3	-1.31119	-1.31052	1.2519	1.4327	7.59347
1964:4	-1.31362	-1.31258	1.30373	1.42863	7.59775
1965:1	-1.30793	-1.30657	1.35841	1.4358	7.61918
1965:2	-1.30175	-1.30108	1.35403	1.43675	7.63356
1965:3	-1.29592	-1.2955	1.35248	1.44621	7.64936
1965:4	-1.29394	-1.29344	1.4272	1.49806	7.67211
1966:1	-1.29519	-1.29451	1.52823	1.56235	7.6917
1966:2	-1.29038	-1.29089	1.52323	1.56444	7.6943
1966:3	-1.27007	-1.26971	1.618	1.63705	7.7045
1966:4	-1.25842	-1.2588	1.65058	1.61004	7.7094
1967:1	-1.25263	-1.25384	1.50696	1.52235	7.71503
1967:2	-1.25705	-1.25819	1.29746	1.57277	7.72099
1967:3	-1.27082	-1.27253	1.45861	1.65766	7.7353
1967:4	-1.27302	-1.27373	1.55878	1.73042	7.74092
1968:1	-1.26666	-1.26646	1.61939	1.72935	7.75246
1968:2	-1.26919	-1.26792	1.70838	1.74798	7.76934
1968:3	-1.27012	-1.27003	1.64808	1.69745	7.77708
1968:4	-1.27253	-1.27148	1.72044	1.75267	7.77611
1969:1	-1.27502	-1.27295	1.80714	1.82083	7.79268
1969:2	-1.2677	-1.26738	1.82406	1.84893	7.79342
1969:3	-1.25224	-1.25295	1.94919	1.92527	7.79799
1969:4	-1.24053	-1.24028	1.99511	1.98746	7.78734
1970:1	-1.23704	-1.23529	1.97547	1.99742	7.7868
1970:2	-1.22588	-1.22632	1.89867	2.04291	7.78593
1970:3	-1.22078	-1.22207	1.8453	2.00955	7.79803
1970:4	-1.22328	-1.22335	1.67766	1.92469	7.78896
1971:1	-1.22377	-1.22197	1.34547	1.79459	7.81545
1971:2	-1.22733	-1.22774	1.44692	1.8321	7.81537

Period	*K*	*KSA*	*R3M*	*R10Y*	*RGNP*
1971:3	-1.22089	-1.22173	1.61144	1.86918	7.82048
1971:4	-1.21619	-1.21626	1.4422	1.77326	7.82044
1972:1	-1.22181	-1.21992	1.2346	1.79724	7.84212
1972:2	-1.22345	-1.22436	1.32707	1.81531	7.86138
1972:3	-1.22676	-1.22697	1.43983	1.83896	7.87173
1972:4	-1.22592	-1.22641	1.58166	1.85207	7.89032
1973:1	-1.22619	-1.22516	1.74047	1.88813	7.91352
1973:2	-1.21215	-1.21298	1.88752	1.91795	7.91608
1973:3	-1.20443	-1.20388	2.11902	1.97547	7.91509
1973:4	-1.19443	-1.19554	2.0149	1.90999	7.924
1974:1	-1.18328	-1.18323	2.03038	1.95345	7.91841
1974:2	-1.16377	-1.1643	2.09839	2.02062	7.92124
1974:3	-1.14907	-1.14868	2.10291	2.07481	7.90813
1974:4	-1.12812	-1.12907	1.99606	2.03732	7.8993
1975:1	-1.10741	-1.10677	1.7492	2.02022	7.87956
1975:2	-1.0999	-1.09994	1.6851	2.08567	7.88968
1975:3	-1.09249	-1.09229	1.8453	2.11589	7.90651
1975:4	-1.07688	-1.07726	1.72758	2.08729	7.92034
1976:1	-1.06438	-1.06384	1.5927	2.04808	7.93894
1976:2	-1.04871	-1.04947	1.64035	2.05066	7.94339
1976:3	-1.0362	-1.03583	1.639	2.04511	7.94754
1976:4	-1.03508	-1.03516	1.5418	1.97269	7.95746
1977:1	-1.03776	-1.03677	1.53256	1.99511	7.97109
1977:2	-1.03149	-1.03236	1.57691	1.99742	7.98708
1977:3	-1.02216	-1.02258	1.7042	1.99565	8.00697
1977:4	-1.01837	-1.01829	1.80993	2.02775	8.0044
1978:1	-1.01113	-1.0108	1.8552	2.08069	8.01318
1978:2	-1.01007	-1.01096	1.86826	2.11866	8.04427
1978:3	-1.00737	-1.00767	1.98965	2.13889	8.05281
1978:4	-0.99573	-0.99482	2.14827	2.17702	8.06514
1979:1	-0.98062	-0.97907	2.2389	2.20904	8.06517
1979:2	-0.97524	-0.98476	2.23826	2.2097	8.06423
1979:3	-0.98611	-0.98243	2.26934	2.2086	8.07322
1979:4	-0.97149	-0.96555	2.47174	2.34631	8.07131
1980:1	-0.95585	-0.9558	2.59174	2.48382	8.08129
1980:2	-0.90995	-0.91982	2.26353	2.34918	8.05738
1980:3	-0.93938	-0.93518	2.21408	2.39361	8.05804
1980:4	-0.94294	-0.93662	2.61102	2.51955	8.07066

Period	*K*	*KSA*	*R3M*	*R10Y*	*RGNP*
1981:1	-0.93048	-0.93088	2.66653	2.56187	8.08982
1981:2	-0.929	-0.93912	2.70183	2.62104	8.08647
1981:3	-0.93421	-0.92874	2.71158	2.6978	8.09089
1981:4	-0.93229	-0.92743	2.46385	2.64525	8.07683
1982:1	-0.92987	-0.93196	2.55046	2.65977	8.06161
1982:2	-0.90719	-0.91476	2.51931	2.63404	8.0646
1982:3	-0.91043	-0.90575	2.23184	2.57391	8.05658
1982:4	-0.93609	-0.93308	2.06775	2.36715	8.05811
1983:1	-0.93585	-0.93792	2.09273	2.35736	8.06671
1983:2	-0.93894	-0.94272	2.12787	2.35546	8.08896
1983:3	-0.94946	-0.94585	2.21266	2.45333	8.10361
1983:4	-0.93548	-0.93375	2.17475	2.45848	8.12121
1984:1	-0.92676	-0.92877	2.21594	2.48014	8.14662
1984:2	-0.92447	-0.92584	2.28208	2.58022	8.15995
1984:3	-0.91616	-0.91419	2.33408	2.55467	8.16639
1984:4	-0.91255	-0.91199	2.17509	2.46326	8.17053
1985:1	-0.92642	-0.92757	2.10206	2.44954	8.18242
1985:2	-0.9375	-0.93724	2.00955	2.38075	8.18847
1985:3	-0.95643	-0.95564	1.96108	2.33573	8.19858
1985:4	-0.97057	-0.9695	1.96949	2.27829	8.20587
1986:1	-0.97807	-0.97883	1.93109	2.14675	8.22129
1986:2	-1.00655	-1.00604	1.81482	2.02854	8.21922
1986:3	-1.03834	-1.03914	1.70892	1.98883	8.22183
1986:4	-1.07117	-1.06976	1.67766	1.98279	8.22542
1987:1	-1.08421	-1.08556	1.71145	1.97311	8.23661
1987:2	-1.08205	-1.08176	1.73289	2.12142	8.24879
1987:3	-1.05939	-1.06047	1.7989	2.18346	8.25979
1987:4	-1.04268	-1.04002	1.76866	2.2108	8.27461
1988:1	-1.0213	-1.02283	1.74449	2.13025	8.28301
1988:2	-1.01601	-1.01598	1.82616	2.18717	8.29034
1988:3	-1.00832	-1.0097	1.94734	2.20827	8.2964
1988:4	-0.99413	-0.99109	2.04472	2.19243	8.30237

Notes:

The definitions of the variables, all transformed to natural logarithms, are as follows: K is the ratio of currency to total checkable deposits; KSA is seasonally adjusted K; $R3M$ is nominal 3 month Treasury bill rate; $R10Y$ is nominal yield on 10-year Government securities; $RGNP$ is real GNP denoted also as q.

Table D.2

Data Used in Chapter 2: 1953:1 to 1988:4

Period	$(M1/P)$	$(M2/P)$	(MB/P)	$(NM1M2/P)$
1953:1	6.18533	6.73401	5.0533	5.87195
1953:2	6.1876	6.72621	5.05057	5.85018
1953:3	6.18533	6.73401	5.0533	5.87195
1953:4	6.19516	6.75341	5.06267	5.90429
1954:1	6.18294	6.75076	5.05218	5.91429
1954:2	6.18019	6.76099	5.04592	5.94126
1954:3	6.1903	6.77888	5.04347	5.96895
1954:4	6.1933	6.78676	5.04726	5.98288
1955:1	6.1936	6.79142	5.04005	5.99288
1955:2	6.1881	6.79237	5.02488	6.00167
1955:3	6.18522	6.79614	5.01995	6.01339
1955:4	6.17991	6.79902	5.01908	6.02592
1956:1	6.17647	6.80132	5.01584	6.03487
1956:2	6.17109	6.80478	5.00944	6.04841
1956:3	6.16108	6.80449	5.00352	6.05901
1956:4	6.15822	6.8089	4.99168	6.07139
1957:1	6.14652	6.80804	4.97848	6.08223
1957:2	6.14337	6.81614	4.97978	6.10219
1957:3	6.13301	6.81713	4.97342	6.11486
1957:4	6.12481	6.82217	4.97	6.13322
1958:1	6.11648	6.8294	4.96161	6.15563
1958:2	6.12465	6.851	4.97237	6.18998
1958:3	6.13154	6.86756	4.97133	6.21553
1958:4	6.13762	6.87708	4.96695	6.2282
1959:1	6.13444	6.85983	4.96703	6.19791
1959:2	6.13617	6.86805	4.96654	6.2122
1959:3	6.1367	6.87476	4.96305	6.22458
1959:4	6.12795	6.87858	4.96305	6.23978
1960:1	6.11534	6.87396	4.95405	6.24227
1960:2	6.1132	6.88186	4.95023	6.25892
1960:3	6.11923	6.89629	4.95385	6.28056
1960:4	6.119	6.91098	4.96291	6.30776
1961:1	6.12372	6.92798	4.96368	6.33482
1961:2	6.1248	6.93982	4.95799	6.3553
1961:3	6.12423	6.95006	4.95983	6.37403
1961:4	6.13325	6.96771	4.97315	6.39835

Period	$(M1/P)$	$(M2/P)$	(MB/P)	$(NM1M2/P)$
1962:1	6.13018	6.97878	4.96948	6.42014
1962:2	6.1305	6.99304	4.97409	6.44471
1962:3	6.12936	7.00866	4.98057	6.47238
1962:4	6.1259	7.01922	4.97979	6.49278
1963:1	6.13271	7.0381	4.98873	6.51994
1963:2	6.14207	7.05951	5.00062	6.54946
1963:3	6.14913	7.07653	5.01064	6.57303
1963:4	6.15282	7.08999	5.01539	6.59285
1964:1	6.15605	7.10324	5.02515	6.6125
1964:2	6.1601	7.11698	5.03539	6.63233
1964:3	6.17042	7.13308	5.04376	6.65201
1964:4	6.18019	7.15096	5.05496	6.67487
1965:1	6.17565	7.1585	5.05381	6.6897
1965:2	6.17865	7.1718	5.06282	6.70911
1965:3	6.18101	7.18192	5.06764	6.72378
1965:4	6.19327	7.19851	5.07774	6.74288
1966:1	6.19864	7.20553	5.08239	6.75085
1966:2	6.20091	7.20857	5.08625	6.75432
1966:3	6.18922	7.206	5.08709	6.75696
1966:4	6.18061	7.20497	5.08102	6.7602
1967:1	6.18485	7.21506	5.09046	6.77355
1967:2	6.1988	7.23881	5.10302	6.80271
1967:3	6.2122	7.25769	5.11043	6.82457
1967:4	6.21717	7.26712	5.11873	6.8364
1968:1	6.21155	7.26487	5.11437	6.83597
1968:2	6.2178	7.27085	5.11707	6.84181
1968:3	6.2265	7.27996	5.12597	6.85112
1968:4	6.23212	7.28643	5.12993	6.85805
1969:1	6.23692	7.29022	5.13228	6.86131
1969:2	6.23234	7.28677	5.12708	6.85846
1969:3	6.22157	7.27843	5.11898	6.85142
1969:4	6.21723	7.27614	5.11698	6.85022
1970:1	6.21121	7.26398	5.11016	6.83478
1970:2	6.20373	7.25682	5.11025	6.82779
1970:3	6.2097	7.27007	5.12166	6.84492
1970:4	6.21554	7.28498	5.12718	6.86461
1971:1	6.21645	7.30039	5.1338	6.88751
1971:2	6.21867	7.32133	5.13349	6.91788

Period	$(M1/P)$	$(M2/P)$	(MB/P)	$(NM1M2/P)$
1971:3	6.22455	7.33855	5.14292	6.94069
1971:4	6.22048	7.35247	5.14128	6.96329
1972:1	6.22935	7.3711	5.14865	6.98652
1972:2	6.23861	7.39033	5.16096	7.0104
1972:3	6.24609	7.41002	5.16488	7.03569
1972:4	6.25778	7.42968	5.1767	7.05894
1973:1	6.26287	7.43932	5.18327	7.07062
1973:2	6.25396	7.43542	5.18193	7.06894
1973:3	6.24546	7.43023	5.17878	7.06521
1973:4	6.2342	7.4201	5.17146	7.05558
1974:1	6.23721	7.4253	5.17908	7.06174
1974:2	6.22512	7.41688	5.18065	7.05493
1974:3	6.20089	7.39435	5.16513	7.03313
1974:4	6.1859	7.38208	5.15876	7.02204
1975:1	6.16741	7.3757	5.15007	7.02087
1975:2	6.16648	7.39615	5.15383	7.05028
1975:3	6.16346	7.40996	5.15175	7.07098
1975:4	6.15355	7.41622	5.14952	7.08368
1976:1	6.15541	7.43584	5.15447	7.11022
1976:2	6.15822	7.45355	5.1638	7.1336
1976:3	6.1547	7.46629	5.16553	7.15241
1976:4	6.15674	7.48559	5.16737	7.178
1977:1	6.16299	7.50184	5.16855	7.19783
1977:2	6.16066	7.50915	5.16813	7.20854
1977:3	6.16598	7.52125	5.17897	7.22301
1977:4	6.16999	7.52588	5.18226	7.22785
1978:1	6.17452	7.52974	5.19306	7.23148
1978:2	6.17229	7.52309	5.19026	7.22329
1978:3	6.17449	7.52488	5.19408	7.22493
1978:4	6.17214	7.52591	5.19529	7.22715
1979:1	6.16108	7.51944	5.19018	7.22227
1979:2	6.15703	7.51851	5.18598	7.22242
1979:3	6.17058	7.52491	5.18954	7.22634
1979:4	6.16154	7.51976	5.19112	7.22254
1980:1	6.15139	7.51497	5.18829	7.2196
1980:2	6.11344	7.50163	5.17876	7.21458
1980:3	6.13891	7.5163	5.18457	7.22564
1980:4	6.13926	7.51125	5.17765	7.21876

Period	$(M1/P)$	$(M2/P)$	(MB/P)	$(NM1M2/P)$
1981:1	6.11922	7.50145	5.15698	7.21241
1981:2	6.11989	7.5088	5.16005	7.22199
1981:3	6.11396	7.51007	5.14712	7.22564
1981:4	6.10626	7.51584	5.13483	7.2358
1982:1	6.10993	7.52327	5.14153	7.24444
1982:2	6.10282	7.52764	5.14436	7.25248
1982:3	6.10991	7.53792	5.14494	7.26376
1982:4	6.13991	7.55328	5.15854	7.27446
1983:1	6.15622	7.59502	5.1795	7.32424
1983:2	6.17686	7.613	5.19935	7.34139
1983:3	6.19705	7.62325	5.21166	7.34852
1983:4	6.20143	7.6321	5.22058	7.35878
1984:1	6.20536	7.64039	5.23334	7.36844
1984:2	6.21417	7.65088	5.24291	7.37945
1984:3	6.21738	7.65818	5.24955	7.38803
1984:4	6.21943	7.67284	5.25248	7.40656
1985:1	6.23857	7.69602	5.26401	7.43097
1985:2	6.25536	7.70293	5.27391	7.43487
1985:3	6.28452	7.72053	5.29291	7.44888
1985:4	6.30412	7.72911	5.30601	7.45401
1986:1	6.32395	7.74097	5.32002	7.46333
1986:2	6.35337	7.75663	5.33525	7.47454
1986:3	6.38367	7.77245	5.35	7.48559
1986:4	6.4221	7.79082	5.37026	7.49721
1987:1	6.44551	7.79732	5.38895	7.49788
1987:2	6.45292	7.79411	5.39701	7.49092
1987:3	6.44676	7.79499	5.40088	7.49429
1987:4	6.45325	7.80134	5.41239	7.50059
1988:1	6.45702	7.81232	5.42874	7.51409
1988:2	6.45957	7.8161	5.4334	7.5183
1988:3	6.46084	7.81404	5.43824	7.51508
1988:4	6.45367	7.81007	5.43585	7.51222

Notes:

The definitions of the variables, all transformed to natural logarithms, are as follows: $(M1/P)$ is Real $M1$; $(M2/P)$ is real $M2$; (MB/P) is real adjusted monetary base and $(NM1M2/P)$ is real non-$M1$ component of $M2$.

Table D.3

Data Used in Chapter 3: 1956:4 to 1991:2

Period	*C*	*I*	*W*	*DD*681	*DD*792	*DD*883
1966:4	10.4831	10.5821	12.9481			
1967:1	10.4893	10.5800	12.9895	0	0	0
1967:2	10.5022	10.5990	13.0115	0	0	0
1967:3	10.5240	10.6262	13.0411	0	0	0
1967:4	10.5329	10.6145	13.0357	0	0	0
1968:1	10.5586	10.6307	13.0518	0	0	0
1968:2	10.5190	10.6316	13.0839	1	0	0
1968:3	10.5381	10.6132	13.1120	-1	0	0
1968:4	10.5422	10.6141	13.1183	0	0	0
1969:1	10.5361	10.6263	13.1144	0	0	0
1969:2	10.5462	10.6366	13.1009	0	0	0
1969:3	10.5459	10.6313	13.0882	0	0	0
1969:4	10.5552	10.6324	13.0402	0	0	0
1970:1	10.5548	10.6361	13.0391	0	0	0
1970:2	10.5710	10.6795	13.0417	0	0	0
1970:3	10.5861	10.6813	13.0261	0	0	0
1970:4	10.5864	10.6801	13.0032	0	0	0
1971:1	10.5802	10.6533	13.0364	0	0	0
1971:2	10.6006	10.6839	13.0461	0	0	0
1971:3	10.6168	10.6889	13.0850	0	0	0
1971:4	10.6275	10.7025	13.1107	0	0	0
1972:1	10.6414	10.7212	13.1241	0	0	0
1972:2	10.6629	10.7818	13.1605	0	0	0
1972:3	10.6758	10.7641	13.1748	0	0	0
1972:4	10.6881	10.7841	13.1612	0	0	0
1973:1	10.7240	10.8045	13.1050	0	0	0
1973:2	10.7143	10.8230	13.1082	0	0	0
1973:3	10.7222	10.8319	13.1059	0	0	0
1973:4	10.7156	10.8380	13.0140	0	0	0
1974:1	10.6964	10.8097	12.9301	0	0	0
1974:2	10.6990	10.7928	12.8427	0	0	0
1974:3	10.7081	10.8310	12.7710	0	0	0
1974:4	10.7142	10.8328	12.7281	0	0	0
1975:1	10.7078	10.8527	12.7692	0	0	0
1975:2	10.7073	10.8089	12.7492	0	0	0
1975:3	10.6954	10.8202	12.7664	0	0	0

Period	*C*	*I*	*W*	*DD*681	*DD*792	*DD*883
1975:4	10.6910	10.8069	12.7554	0	0	0
1976:1	10.6967	10.8196	12.7605	0	0	0
1976:2	10.7015	10.8046	12.7471	0	0	0
1976:3	10.7083	10.8372	12.7238	0	0	0
1976:4	10.7127	10.8123	12.7156	0	0	0
1977:1	10.6922	10.7842	12.7555	0	0	0
1977:2	10.6874	10.7713	12.7517	0	0	0
1977:3	10.6989	10.7904	12.8018	0	0	0
1977:4	10.7224	10.8369	12.8388	0	0	0
1978:1	10.7452	10.8333	12.8438	0	0	0
1978:2	10.7462	10.8635	12.8540	0	0	0
1978:3	10.7663	10.8884	12.8618	0	0	0
1978:4	10.7633	10.8924	12.8491	0	0	0
1979:1	10.7737	10.9017	12.9232	0	0	0
1979:2	10.8282	10.9108	12.9022	0	1	0
1979:3	10.7872	10.9166	12.8737	0	-1	0
1979:4	10.8015	10.9673	12.8467	0	0	0
1980:1	10.8139	10.9324	12.8647	0	0	0
1980:2	10.7909	10.9344	12.8885	0	0	0
1999:3	10.8029	10.9506	12.9183	0	0	0
1980:4	10.7868	10.9465	12.9277	0	0	0
1981:1	10.7979	10.9488	12.9505	0	0	0
1981:2	10.8007	10.9294	12.9615	0	0	0
1981:3	10.8008	10.9248	12.9147	0	0	0
1981:4	10.7991	10.9326	12.9527	0	0	0
1982:1	10.7956	10.9202	12.9641	0	0	0
1982:2	10.8005	10.9373	12.9780	0	0	0
1982:3	10.8160	10.9269	13.0299	0	0	0
1982:4	10.8260	10.9315	13.0604	0	0	0
1983:1	10.8405	10.9399	13.1031	0	0	0
1983:2	10.8482	10.9599	13.1577	0	0	0
1983:3	10.8633	10.9563	13.1504	0	0	0
1983:4	10.8633	10.9637	13.1805	0	0	0
1984:1	10.8615	10.9703	13.2245	0	0	0
1984:2	10.8732	10.9778	13.1852	0	0	0
1984:3	10.8649	10.9801	13.2298	0	0	0
1984:4	10.8793	10.9942	13.2849	0	0	0
1985:1	10.8909	10.9840	13.2999	0	0	0

Period	C	I	W	$DD681$	$DD792$	$DD883$
1985:2	10.8938	11.0120	13.2904	0	0	0
1985:3	10.9116	11.0120	13.3140	0	0	0
1985:4	10.9202	11.0237	13.3606	0	0	0
1986:1	10.9409	11.0300	13.4574	0	0	0
1986:2	10.9663	11.0624	13.4655	0	0	0
1986:3	10.9700	11.0556	13.4371	0	0	0
1986:4	10.9808	11.0644	13.5020	0	0	0
1987:1	10.9878	11.0618	13.5914	0	0	0
1987:2	11.0048	11.0839	13.6804	0	0	0
1987:3	11.0272	11.0944	13.7131	0	0	0
1987:4	11.0420	11.1095	13.5633	0	0	0
1988:1	11.0701	11.1116	13.5814	0	0	0
1988:2	11.0751	11.1413	13.6171	0	0	0
1988:3	11.0964	11.1507	13.6201	0	0	1
1988:4	11.1069	11.1680	13.6460	0	0	1
1989:1	11.1123	11.1713	13.6731	0	0	1
1989:2	11.1231	11.2032	13.6884	0	0	1
1989:3	11.1223	11.2009	13.7211	0	0	1
1989:4	11.1303	11.2064	13.7686	0	0	1
1990:1	11.1307	11.2160	13.6833	0	0	1
1990:2	11.1389	11.2147	13.7130	0	0	1
1990:3	11.1325	11.2286	13.6225	0	0	1
1990:4	11.1261	11.2352	13.6957	0	0	1
1991:1	11.1232	11.2189	13.7723	0	0	1
1991:2	11.1220	11.2276	13.7424	0	0	1

Notes:

C = Real consumption expenditure, I = real income and W = real wealth. The other three variables are dummy variables. For further details see the explanation given earlier in this appendix.

Table D.4

Data on Real Aggregate Output in Chapter 4: 1870 to 1986

Period	*Australia*	*Canada*	*Denmark*	*Finland*	*France*	*Germany*
1870	1.4157	1.6415	2.8758	*NA*	5.7355	4.0812
1871	1.3758	1.6709	2.8789	*NA*	5.7713	4.0746
1872	1.4800	1.4972	2.9328	*NA*	5.8447	4.1418
1873	1.5809	1.5904	2.9270	*NA*	5.7810	4.1842
1874	1.6135	1.7806	2.9559	*NA*	5.9046	4.2558
1875	1.7140	1.7453	2.9728	*NA*	5.9464	4.2613
1876	1.7140	1.7543	2.9922	*NA*	5.9063	4.2558
1877	1.7510	1.7806	2.9644	*NA*	5.9282	4.2502
1878	1.8417	1.7631	3.0032	*NA*	5.9080	4.2964
1879	1.8594	1.9217	3.0353	*NA*	5.8768	4.2722
1880	1.9107	2.0285	3.0587	*NA*	5.9349	4.2640
1881	1.9797	2.1553	3.0689	*NA*	5.9643	4.2884
1882	1.9202	2.1493	3.1039	*NA*	6.0007	4.3043
1883	2.0587	2.1312	3.1377	*NA*	5.9803	4.3581
1884	2.0669	2.2271	3.1425	*NA*	5.9562	4.3827
1885	2.1301	2.3296	3.1495	*NA*	5.9643	4.4068
1886	2.1396	2.2271	3.1887	*NA*	6.0007	4.4139
1887	2.2740	2.2271	3.2241	*NA*	5.9803	4.4532
1888	2.2489	2.2783	3.2307	*NA*	5.9562	4.4932
1889	2.3304	2.2544	3.2436	*NA*	5.9643	4.5211
1890	2.2954	2.2783	3.3017	*NA*	5.9771	4.5524
1891	2.3703	2.4317	3.3216	*NA*	5.9803	4.5503
1892	2.2387	2.4089	3.3451	*NA*	5.9739	4.5907
1893	2.1822	2.3808	3.3642	*NA*	5.9897	4.6389
1894	2.2161	2.3760	3.3848	*NA*	6.0176	4.6631
1895	2.1565	2.4272	3.4390	*NA*	6.0357	4.7098
1896	2.2300	2.5203	3.4752	*NA*	6.0638	4.7443
1897	2.1731	2.3784	3.4987	*NA*	6.0387	4.7727
1898	2.3178	2.5488	3.5151	*NA*	6.1080	4.8146
1899	2.3194	2.5057	3.5566	*NA*	6.0869	4.8503
1900	2.3763	2.5508	3.5904	2.7025	6.1177	4.8921
1901	2.3474	2.6444	3.6318	2.6776	6.0996	4.8684
1902	2.3612	2.7149	3.6547	2.6459	6.1530	4.8921
1903	2.3822	2.7434	3.7125	2.7310	6.2037	4.9462
1904	2.4273	2.7614	3.7337	2.7723	6.2213	4.9855
1905	2.4400	2.8350	3.7505	2.7844	6.1871	5.0066

Period	*Australia*	*Canada*	*Denmark*	*Finland*	*France*	*Germany*
1906	2.5298	2.8907	3.7785	2.8211	6.1871	5.0362
1907	2.5551	2.8950	3.8153	2.8809	6.2213	5.0795
1908	2.5336	2.9367	3.8462	2.8785	6.2687	5.0963
1909	2.5810	3.0438	3.8841	2.8858	6.2532	5.1164
1910	2.6527	3.0632	3.9140	2.9131	6.2687	5.1520
1911	2.6872	3.1461	3.9662	2.9522	6.2840	5.1852
1912	2.7280	3.1647	3.9662	3.0323	6.2991	5.2279
1913	2.7782	3.2459	4.0028	3.0721	6.3151	5.2719
1914	2.7206	3.1830	4.0640	2.9985	6.3151	5.1117
1915	2.7110	3.2134	3.9918	2.9778	6.3584	5.0599
1916	2.7488	3.2569	4.0334	3.0271	6.4407	5.0698
1917	2.7238	3.2784	3.9724	2.8614	6.3831	5.0710
1918	2.7269	3.2871	3.9388	2.6242	6.3003	5.0734
1919	2.7100	3.3330	4.0602	2.7884	6.2568	4.9475
1920	2.7478	3.2823	4.1063	2.9522	6.2263	5.0311
1921	2.8356	3.1872	4.0770	2.9887	6.1715	5.1383
1922	2.8781	3.2647	4.1734	3.1017	6.1557	5.2227
1923	2.9136	3.3257	4.2736	3.1592	6.2398	5.0374
1924	2.9647	3.3238	4.2767	3.2136	6.2238	5.1950
1925	2.9812	3.3664	4.2536	3.2726	6.3692	5.3014
1926	2.9869	3.4505	4.3103	3.3087	6.4215	5.3292
1927	3.0070	3.5408	4.3300	3.3898	6.5195	5.4246
1928	2.9885	3.6285	4.3633	3.4228	6.5305	5.4675
1929	2.9861	3.6326	4.4281	3.4171	6.5424	5.4633
1930	2.9446	3.5888	4.4859	3.4007	6.5305	5.4492
1931	2.9031	3.4529	4.4969	3.3558	6.5865	5.3699
1932	2.9404	3.3430	4.4704	3.3737	6.6702	5.2917
1933	2.9877	3.2745	4.5018	3.4416	6.6396	5.3525
1934	3.0173	3.3892	4.5317	3.5601	6.5977	5.4399
1935	3.0528	3.4643	4.5539	3.5837	6.5540	5.5125
1936	3.0938	3.5075	4.5783	3.6515	6.5977	5.5969
1937	3.1408	3.6029	4.6022	3.7456	6.5977	5.6998
1938	3.1512	3.6113	4.6261	3.7527	6.5647	5.7972
1939	3.1580	3.6829	4.6727	3.7250	6.5761	5.8746
1940	3.2210	3.8146	4.5217	3.5980	6.6297	5.8816
1941	3.3275	3.9490	4.4176	3.6571	6.6297	5.9433
1942	3.4363	4.1193	4.4398	3.7018	6.6702	5.9570
1943	3.4709	4.1590	4.5446	3.7880	6.4780	5.9765

Period	*Australia*	*Canada*	*Denmark*	*Finland*	*France*	*Germany*
1944	3.4358	4.1976	4.6442	3.7692	6.2459	6.0019
1945	3.3847	4.1749	4.5659	3.6092	6.1342	5.6455
1946	3.3481	4.1477	4.7109	3.7312	6.0812	5.0855
1947	3.3721	4.1894	4.7655	3.7953	5.9464	5.2907
1948	3.4347	4.2141	4.7981	3.8583	6.0282	5.4608
1949	3.4980	4.2515	4.8423	3.8998	6.4467	5.6133
1950	3.5763	4.3245	4.9119	3.9600	6.5315	5.7481
1951	3.6339	4.3730	4.9060	4.0492	6.5900	5.8470
1952	3.6540	4.4590	4.9207	4.0831	6.6151	5.9325
1953	3.6522	4.5088	4.9781	4.0842	6.6442	6.0113
1954	3.7141	4.4964	5.0066	4.1707	6.6912	6.0831
1955	3.7661	4.5963	5.0066	4.2436	6.7457	6.1966
1956	3.8201	4.6775	5.0265	4.2631	6.7939	6.2670
1957	3.8450	4.7005	5.0783	4.2784	6.8521	6.3220
1958	3.8609	4.7238	5.1023	4.2774	6.8774	6.3587
1959	3.9466	4.7605	5.1671	4.3476	6.9537	6.4291
1960	3.9989	4.7892	5.2295	4.4423	7.0229	6.5958
1961	3.9910	4.7983	5.2883	4.5155	7.0766	6.6383
1962	4.0565	4.8691	5.3414	4.5449	7.1413	6.6845
1963	4.1205	4.9185	5.3476	4.5773	7.1937	6.7115
1964	4.1787	4.9836	5.4333	4.6279	7.2563	6.7750
1965	4.2360	5.0466	5.4806	4.6803	7.3079	6.8273
1966	4.2583	5.1126	5.5318	4.7032	7.3588	6.8566
1967	4.3124	5.1411	5.5680	4.7247	7.4049	6.8557
1968	4.3810	5.1952	5.6051	4.7475	7.4466	6.9117
1969	4.4610	5.2486	5.6685	4.8315	7.5142	6.9847
1970	4.5208	5.2735	5.6907	4.9112	7.6153	7.0335
1971	4.5774	5.3293	5.7150	4.9323	7.6620	7.0630
1972	4.6131	5.3868	5.7680	5.0053	7.7053	7.1041
1973	4.6738	5.4613	5.8159	5.0702	7.7583	7.1499
1974	4.6867	5.5037	5.8063	5.1005	7.7890	7.1515
1975	4.7113	5.5290	5.7997	5.1120	7.7861	7.1373
1976	4.7484	5.5854	5.8625	5.1144	7.8276	7.1914
1977	4.7562	5.6179	5.8786	5.1156	7.8594	7.2174
1978	4.7916	5.6588	5.8933	5.1370	7.8925	7.2499
1979	4.8259	5.6954	5.9283	5.2078	7.9241	7.2889
1980	4.8450	5.7108	5.9237	5.2606	7.9402	7.3032
1981	4.8798	5.7401	5.9149	5.2766	7.9519	7.3032

Period	Australia	Canada	Denmark	Finland	France	Germany
1982	4.8851	5.7054	5.9446	5.3122	7.9773	7.2937
1983	4.8888	5.7420	5.9695	5.3414	7.9841	7.3125
1984	4.9544	5.8012	6.0122	5.3734	7.9973	7.3447
1985	4.9986	5.8433	6.0533	5.4076	8.0137	7.3652
1986	5.0120	5.8727	6.0861	5.4315	8.0343	7.3890

Table D.5

Data on Real Aggregate Output in Chapter 4: 1870 to 1986

Period	Italy	Norway	Sweden	U.K.	U.S.A.
1870	10.2495	2.5382	3.0781	3.4137	6.6172
1871	10.2420	2.5529	3.1131	3.4661	6.6788
1872	10.2325	2.6141	3.1737	3.4682	6.7522
1873	10.2716	2.6344	3.2281	3.4913	6.7672
1874	10.2716	2.6674	3.2392	3.5078	6.7622
1875	10.2987	2.6973	3.2197	3.5320	6.7966
1876	10.2880	2.7263	3.2797	3.5419	6.8392
1877	10.2862	2.7303	3.2797	3.5517	6.9020
1878	10.2898	2.6994	3.2770	3.5556	6.9611
1879	10.3005	2.7119	3.3362	3.5517	7.0168
1880	10.3163	2.7425	3.3437	3.5975	7.0879
1881	10.2661	2.7505	3.3585	3.6323	7.1303
1882	10.3163	2.7465	3.3609	3.6608	7.1710
1883	10.3111	2.7425	3.4132	3.6677	7.1972
1884	10.3233	2.7505	3.4132	3.6695	7.2165
1885	10.3337	2.7465	3.4270	3.6642	7.2353
1886	10.3524	2.7425	3.4383	3.6798	7.3016
1887	10.3591	2.7623	3.4247	3.7186	7.3416
1888	10.3473	2.7741	3.4651	3.7622	7.3665
1889	10.3163	2.7799	3.4760	3.8146	7.4223
1890	10.3839	2.7914	3.5060	3.8191	7.4922
1891	10.3937	2.8345	3.5453	3.8191	7.5346
1892	10.3541	2.8687	3.5594	3.7949	7.6272
1893	10.3872	2.8949	3.5870	3.7949	7.5776
1894	10.3757	2.9035	3.6139	3.8601	7.5484
1895	10.3969	2.9255	3.6709	3.8912	7.6624

Period	*Italy*	*Norway*	*Sweden*	*U.K.*	*U.S.A.*
1896	10.4129	2.9518	3.7043	3.9321	7.6418
1897	10.3724	2.9551	3.7450	3.9454	7.7330
1898	10.4349	2.9648	3.7713	3.9930	7.7536
1899	10.4488	2.9949	3.7905	4.0336	7.8406
1900	10.5154	3.0436	3.8156	4.0263	7.8674
1901	10.5657	3.0466	3.8047	4.0263	7.9741
1902	10.5711	3.0743	3.8415	4.0515	7.9845
1903	10.5846	3.0915	3.8916	4.0408	8.0320
1904	10.5859	3.1180	3.9226	4.0468	8.0192
1905	10.6201	3.1385	3.9419	4.0761	8.0906
1906	10.6405	3.1344	4.0269	4.1091	8.1997
1907	10.7368	3.1331	4.0674	4.1279	8.2150
1908	10.7357	3.1452	4.0710	4.0864	8.1296
1909	10.7959	3.1808	4.0734	4.1091	8.2449
1910	10.7437	3.2177	4.1337	4.1399	8.2551
1911	10.8035	3.2497	4.1757	4.1688	8.2872
1912	10.8268	3.2759	4.2127	4.1834	8.3329
1913	10.8749	3.3106	4.2525	4.2211	8.3716
1914	10.8649	3.3409	4.2545	4.2311	8.2915
1915	10.9775	3.3851	4.2374	4.3082	8.3193
1916	11.0779	3.4396	4.2918	4.3300	8.4486
1917	11.1100	3.4614	4.1746	4.3389	8.4233
1918	11.1179	3.5035	4.1692	4.3451	8.5097
1919	10.9501	3.5349	4.2169	4.2301	8.5183
1920	10.8749	3.4396	4.2801	4.1677	8.5088
1921	10.8547	3.4019	4.3098	4.0830	8.4859
1922	10.9132	3.5583	4.3641	4.1334	8.5397
1923	10.9684	3.6194	4.4087	4.1645	8.6636
1924	10.9865	3.5331	4.4290	4.2050	8.6937
1925	11.0539	3.6434	4.5180	4.2526	8.7166
1926	11.0623	3.6652	4.5797	4.2151	8.7798
1927	11.0371	3.6604	4.6234	4.2925	8.7897
1928	11.1100	3.7203	4.6262	4.3045	8.8008
1929	11.1404	3.7419	4.7004	4.3335	8.8602
1930	11.0860	3.7797	4.7563	4.3264	8.7611
1931	11.0779	3.8119	4.6837	4.2737	8.6733
1932	11.1021	3.9008	4.6593	4.2813	8.5297
1933	11.1021	3.9720	4.6824	4.3100	8.5094

Period	*Italy*	*Norway*	*Sweden*	*U.K.*	*U.S.A.*
1934	11.1021	3.8913	4.7430	4.3738	8.5841
1935	11.1919	3.9560	4.7989	4.4117	8.6616
1936	11.1919	3.9796	4.8607	4.4562	8.7951
1937	11.2541	4.0112	4.8763	4.4904	8.8429
1938	11.2670	4.0532	4.9085	4.5025	8.7968
1939	11.3374	4.1126	4.9427	4.5123	8.8728
1940	11.3437	4.1476	4.8981	4.6077	8.9494
1941	11.3311	4.1725	4.8975	4.6949	9.1116
1942	11.3190	4.2189	4.9285	4.7195	9.2845
1943	11.2200	4.1257	4.9452	4.7411	9.4517
1944	11.0121	4.1496	4.9810	4.7011	9.5310
1945	10.7674	4.1100	5.0437	4.6561	9.5127
1946	11.0371	4.0899	5.0945	4.6118	9.2993
1947	11.1985	4.0362	5.1346	4.5968	9.2705
1948	11.2541	4.1501	5.1748	4.6279	9.3084
1949	11.3254	4.2519	5.2267	4.6578	9.3084
1950	11.4032	4.3603	5.2796	4.6932	9.3902
1951	11.4774	4.3813	5.3058	4.7158	9.4880
1952	11.5209	4.4364	5.3216	4.7265	9.5266
1953	11.5927	4.5028	5.3552	4.7698	9.5659
1954	11.6289	4.5375	5.4125	4.8089	9.5525
1955	11.6936	4.5629	5.4411	4.8426	9.6064
1956	11.7393	4.5922	5.4756	4.8590	9.6258
1957	11.7906	4.6308	5.4976	4.8783	9.6421
1958	11.8379	4.6434	5.5263	4.8813	9.6356
1959	11.9016	4.6775	5.5763	4.9214	9.6921
1960	12.0119	4.7283	5.6087	4.9670	9.7135
1961	12.0907	4.8355	5.6605	4.9992	9.7392
1962	12.1511	4.8683	5.6984	5.0093	9.7903
1963	12.2056	4.9053	5.7564	5.0505	9.8304
1964	12.2332	4.9544	5.8222	5.1017	9.8818
1965	12.2458	5.0060	5.8597	5.1251	9.9384
1966	12.3234	5.0428	5.8805	5.1440	9.9956
1967	12.3926	5.1035	5.9135	5.1716	10.0235
1968	12.4561	5.1258	5.9493	5.2133	10.0639
1969	12.5154	5.1699	6.0088	5.2268	10.0883
1970	12.5672	5.1896	6.0700	5.2486	10.0854
1971	12.5831	5.2343	6.0793	5.2745	10.1121

Period	*Italy*	*Norway*	*Sweden*	*U.K.*	*U.S.A.*
1972	12.6149	5.2847	6.1019	5.2963	10.1601
1973	12.6829	5.3250	6.1410	5.3730	10.2080
1974	12.7234	5.3757	6.1723	5.3613	10.2014
1975	12.6863	5.4165	6.1977	5.3542	10.1909
1976	12.7434	5.4823	6.2082	5.3918	10.2378
1977	12.7623	5.5175	6.1920	5.4018	10.2823
1978	12.7888	5.5622	6.2094	5.4407	10.3330
1979	12.8365	5.6113	6.2471	5.4617	10.3536
1980	12.8749	5.6528	6.2636	5.4407	10.3520
1981	12.8864	5.6616	6.2605	5.4311	10.3716
1982	12.8887	5.6647	6.2685	5.4424	10.3466
1983	12.8992	5.7101	6.2925	5.4773	10.3831
1984	12.9305	5.7658	6.3313	5.4976	10.4510
1985	12.9584	5.8174	6.3525	5.5362	10.4863
1986	12.9872	5.8585	6.3641	5.5649	10.5160

Notes:

The data set analyzed here was obtained through the editorial office of the *Journal of Money, Credit and Banking.* The series are annual real GDP except for the United States for which real GNP is used from the National Income and Products Accounts for the period 1929–1986, spliced to Romer's (1989) estimates for the period 1870–1928. For the United Kingdom the series is from Feinstein (1972) for the period 1870–1947 spliced to the International Financial Statistics (IFS) series of the International Monetary Fund for the period 1948–1986. For the remaining countries, the series are from Maddison (1982) spliced to the post-war IFS data. All series are analyzed using a logarithmic transformation.

Table D.6

Data Used in Chapter 5: 1959:1 to 1989:3

Period	*RGNP*	*PGNP*	*ULC*	*GDFCO*	*GDF*	*GDFIM*	*GDFCF*	*GDFCE*
1959: 1	1606.40	1608.30	47.50	36.90	37.40	26.90	32.30	23.10
1959: 2	1637.00	1622.20	47.50	37.40	37.50	27.00	32.20	23.40
1959: 3	1629.50	1636.20	48.70	37.60	37.60	27.10	32.40	23.40
1959: 4	1643.40	1650.30	48.80	37.70	37.80	27.10	32.50	23.80
1960: 1	1671.60	1664.60	49.10	37.80	37.80	27.20	32.40	23.80
1960: 2	1666.80	1679.00	49.60	38.00	38.00	27.40	32.80	23.90
1960: 3	1668.40	1693.50	50.00	38.10	38.10	27.40	32.90	24.10
1960: 4	1654.10	1708.20	50.20	38.20	38.20	27.20	33.20	24.20
1961: 1	1671.30	1722.90	50.10	38.20	38.20	27.20	33.20	24.20
1961: 2	1692.10	1737.80	49.80	38.30	38.20	27.20	33.20	24.20
1961: 3	1716.30	1752.80	49.80	38.40	38.30	27.00	33.10	23.90
1961: 4	1754.90	1768.00	49.50	38.50	38.40	27.00	33.10	24.10
1962: 1	1777.90	1783.30	50.00	38.50	38.40	26.80	33.10	24.00
1962: 2	1796.40	1798.70	50.40	38.70	38.50	26.60	33.30	24.00
1962: 3	1813.10	1814.30	50.10	38.80	38.70	26.60	33.50	24.00
1962: 4	1810.10	1829.90	50.10	39.80	38.80	26.60	33.70	24.30
1963: 1	1834.60	1845.80	50.40	39.90	38.90	26.70	33.90	24.20
1963: 2	1860.00	1861.70	49.90	39.00	38.90	27.00	33.90	24.10
1963: 3	1892.50	1877.80	49.90	39.20	39.00	27.20	34.10	24.10
1963: 4	1906.10	1894.00	50.40	39.30	39.20	27.50	34.20	24.10
1964: 1	1948.70	1910.40	50.10	39.50	39.30	27.60	34.50	24.00
1964: 2	1965.40	1926.90	50.30	39.50	39.40	27.70	34.50	24.00
1964: 3	1985.20	1943.60	50.60	39.60	39.60	27.70	34.60	24.00
1964: 4	1993.70	1960.40	50.90	39.70	39.70	27.70	34.70	24.10
1965: 1	2036.90	1977.40	50.80	39.90	39.90	27.70	34.70	24.30
1965: 2	2066.40	1994.40	50.80	40.00	40.00	27.90	35.20	24.50
1965: 3	2099.30	2012.40	51.00	40.00	40.10	28.10	35.50	24.50
1965: 4	2147.60	2030.50	50.90	40.00	40.20	28.60	35.70	24.60
1966: 1	2190.10	2048.70	51.60	40.20	40.50	28.60	36.50	24.70
1966: 2	2195.80	2067.20	52.60	40.50	40.90	29.10	36.80	24.70
1966: 3	2218.30	2085.80	53.10	40.80	41.20	29.20	37.00	24.90
1966: 4	2229.20	2104.50	53.70	41.20	41.50	29.30	37.20	25.00
1967: 1	2241.80	2123.40	54.30	41.40	41.70	29.40	37.00	25.20
1967: 2	2255.20	2142.50	54.20	41.60	41.90	29.40	37.10	25.30
1967: 3	2287.70	2161.80	54.40	41.90	42.20	29.50	37.50	25.40
1967: 4	2300.60	2181.20	54.80	42.30	42.60	29.60	37.80	25.50

Period	*RGNP*	*PGNP*	*ULC*	*GDFCO*	*GDF*	*GDFIM*	*GDFCF*	*GDFCE*
1968: 1	2327.30	2200.90	56.10	42.80	43.00	29.80	38.10	25.60
1968: 2	2366.90	2220.70	56.40	43.30	43.50	30.10	38.50	25.60
1968: 3	2385.30	2240.60	57.20	43.70	43.80	30.20	38.90	25.70
1968: 4	2383.00	2260.80	58.40	44.20	44.40	30.40	39.40	25.80
1969: 1	2416.50	2281.10	59.30	44.60	44.80	30.70	39.70	26.00
1969: 2	2419.80	2301.60	60.40	45.10	45.30	30.90	40.30	26.40
1969: 3	2433.20	2322.30	61.70	45.60	45.80	31.30	41.00	26.40
1969: 4	2423.50	2343.20	62.90	45.10	46.40	32.00	41.80	26.60
1970: 1	2408.60	2364.30	64.60	46.60	46.80	32.40	42.70	26.50
1970: 2	2406.50	2385.50	64.90	47.10	47.10	32.90	42.90	26.90
1970: 3	2435.80	2407.00	65.10	47.60	47.30	34.00	43.20	27.00
1970: 4	2413.80	2428.60	66.10	48.30	47.80	34.50	43.40	27.50
1971: 1	2478.60	2450.50	66.10	48.90	48.30	35.30	43.40	27.70
1971: 2	2478.40	2472.50	67.10	49.40	48.70	35.30	44.10	27.60
1971: 3	2491.10	2494.80	67.70	49.80	49.00	35.80	44.50	28.20
1971: 4	2491.00	2517.20	68.50	50.10	49.30	36.10	44.90	28.50
1972: 1	2545.60	2539.80	69.10	50.60	49.90	37.00	45.60	28.40
1972: 2	2595.10	2562.70	69.30	51.00	50.10	37.60	45.90	28.50
1972: 3	2622.10	2585.70	69.90	51.40	50.40	38.10	46.50	28.90
1972: 4	2671.30	2609.00	70.10	51.70	50.90	38.70	47.30	29.40
1973: 1	2734.00	2632.40	71.00	52.10	51.70	39.60	49.00	29.90
1973: 2	2741.00	2656.10	73.10	52.80	52.60	41.80	51.20	30.60
1973: 3	2738.30	2680.00	74.70	53.30	53.70	43.00	53.90	31.10
1973: 4	2762.80	2704.10	76.40	54.00	54.40	45.50	55.40	33.00
1974: 1	2747.40	2723.70	78.40	54.90	55.50	49.50	57.60	37.30
1974: 2	2755.20	2743.40	81.50	56.10	56.40	53.70	58.50	40.20
1974: 3	2719.30	2763.30	84.40	57.60	57.80	56.40	59.30	41.10
1974: 4	2695.40	2783.30	86.70	58.90	59.30	58.30	61.40	41.20
1975: 1	2642.70	2803.50	88.60	59.90	60.40	59.60	62.10	42.20
1975: 2	2669.60	2823.80	88.40	60.70	61.10	60.30	62.40	43.10
1975: 3	2714.90	2844.20	88.70	61.60	62.30	59.50	64.40	44.90
1975: 4	2752.70	2864.90	90.80	62.60	63.30	59.60	65.40	46.00
1976: 1	2804.40	2885.60	91.90	63.60	64.00	60.10	65.00	46.00
1976: 2	2816.90	2906.50	93.10	64.50	64.70	61.00	65.00	46.10
1976: 3	2828.60	2927.60	95.00	65.50	65.40	61.90	65.50	47.10
1976: 4	2856.80	2948.80	97.00	66.60	66.30	62.20	65.70	48.40
1977: 1	2896.00	2970.10	97.90	67.80	67.20	64.10	66.60	49.50
1977: 2	2942.70	2991.70	99.20	68.90	68.10	65.80	67.70	50.70

Period	*RGNP*	*PGNP*	*ULC*	*GDFCO*	*GDF*	*GDFIM*	*GDFCF*	*GDFCE*
1977: 3	3001.80	3013.30	99.80	69.90	68.70	67.00	68.70	51.20
1977: 4	2994.10	3035.20	102.90	71.00	69.70	67.60	69.50	51.70
1978: 1	3020.50	3057.20	105.30	71.90	70.70	69.00	71.10	52.30
1978: 2	3115.90	3079.30	106.10	73.20	72.00	70.90	74.00	53.30
1978: 3	3142.60	3101.60	108.40	74.50	73.20	72.00	75.70	54.00
1978: 4	3181.60	3124.10	110.80	75.70	74.70	73.30	77.30	55.60
1979: 1	3181.70	3146.70	114.60	76.90	76.30	75.80	80.10	57.80
1979: 2	3178.70	3169.50	118.20	78.30	78.00	78.60	81.70	63.50
1979: 3	3207.40	3192.50	121.40	79.80	79.70	82.30	82.50	70.40
1979: 4	3201.30	3215.60	124.90	81.50	81.30	86.80	84.10	74.80
1980: 1	3233.40	3238.90	127.60	83.50	83.20	92.40	85.20	82.30
1980: 2	3157.00	3258.40	132.30	85.40	85.10	95.30	87.00	86.20
1980: 3	3159.10	3278.00	134.70	87.30	86.90	98.10	89.80	87.60
1980: 4	3199.20	3297.70	137.10	89.30	89.30	100.00	92.60	89.40
1981: 1	3261.10	3317.60	139.30	91.10	91.50	101.90	94.50	95.60
1981: 2	3250.20	3337.50	142.30	92.90	93.10	102.50	95.40	97.90
1981: 3	3264.60	3357.60	145.20	94.90	95.10	100.70	96.80	98.60
1981: 4	3219.00	3377.80	149.40	96.60	96.90	101.00	97.30	100.20
1982: 1	3170.40	3398.20	153.10	97.90	98.20	101.40	99.20	99.60
1982: 2	3179.90	3418.60	154.90	99.20	99.40	100.00	99.90	97.30
1982: 3	3154.50	3439.20	157.70	100.70	100.70	99.30	100.30	100.90
1982: 4	3159.30	3459.90	158.70	102.20	101.70	99.30	100.60	102.20
1983: 1	3186.60	3480.70	158.60	103.50	102.60	98.30	100.70	99.30
1983: 2	3258.30	3501.70	156.70	104.60	103.60	97.40	102.00	100.80
1983: 3	3306.40	3522.70	157.10	105.90	104.60	97.70	102.10	102.00
1983: 4	3365.10	3543.90	158.20	107.10	105.70	97.60	103.20	101.80
1984: 1	3451.70	3565.30	159.10	108.10	106.80	97.70	105.40	102.30
1984: 2	3498.00	3586.70	159.40	109.10	107.80	98.10	105.70	102.70
1984: 3	3520.60	3608.30	161.20	110.30	108.70	97.40	106.70	102.20
1984: 4	3535.20	3630.00	162.30	111.40	109.60	96.80	107.20	102.10
1985: 1	3577.50	3651.90	163.40	112.50	110.60	95.20	107.80	101.60
1985: 2	3599.20	3673.80	163.90	113.60	111.50	95.60	108.20	104.50
1985: 3	3635.80	3695.90	165.10	114.80	112.30	95.30	108.60	104.20
1985: 4	3662.40	3718.20	167.10	116.20	113.20	96.80	109.80	103.50
1986: 1	3721.10	3740.60	166.40	117.20	113.80	95.90	110.30	99.90
1986: 2	3704.60	3763.10	168.50	118.30	114.40	92.50	111.00	91.00
1986: 3	3712.40	3785.70	171.20	119.80	115.30	92.90	113.10	87.60
1986: 4	3733.60	3808.50	173.20	121.20	116.10	94.70	114.40	85.60

Period	*RGNP*	*PGNP*	*ULC*	*GDFCO*	*GDF*	*GDFIM*	*GDFCF*	*GDFCE*
1987: 1	3783.00	3831.40	173.60	122.90	117.40	97.90	115.50	89.40
1987: 2	3823.50	3854.50	173.40	124.50	118.50	100.40	117.10	91.30
1987: 3	3872.80	3877.70	173.90	126.0	119.60	101.90	117.70	93.10
1987: 4	3935.60	3901.00	175.80	127.60	120.80	103.30	118.70	92.90
1988: 1	3974.80	3924.50	175.70	128.70	121.90	104.50	119.40	91.50
1988: 2	4010.70	3948.10	178.70	130.30	123.30	106.10	121.00	92.30
1988: 3	4042.70	3971.90	179.60	131.50	124.90	106.20	123.50	92.80
1988: 4	4069.40	3995.80	181.30	133.30	126.20	107.30	124.90	92.90
1989: 1	4106.80	4019.90	184.10	134.80	127.70	109.50	126.60	94.00
1989: 2	4132.50	4044.10	186.10	134.80	129.30	111.10	129.00	100.60
1989: 3	4162.90	4068.40	187.40	137.20	130.20	109.80	129.90	98.20

Notes:

The variables in columns 2 to 9 are, respectively, *RGNP*, *PGNP*, *ULC*, *GDFCO*, *GDF*, *GDFIM*, *GDFCF* and *GDFCE*. Data on the Nixon price and wage control dummies are not given since they can be easily constructed using the definitions given earlier in this appendix. For further details see the earler discussion in this appendix.

Table D.7

Data Used in Chapter 6: 1956:1 to 1988:4

Period	*M*1	*P*	*GDP*	*R*
1956:1	4681.00000	0.27302	116196.00	2.85333
1956:2	4714.66406	0.27583	117464.00	3.50000
1956:3	4700.00000	0.28221	118336.00	4.06333
1956:4	4654.00000	0.27927	122060.00	4.52000
1957:1	4595.00000	0.28136	121508.00	5.20667
1957:2	4608.66406	0.28350	121300.00	5.32000
1957:3	4562.66406	0.28595	122160.00	5.32000
1957:4	4638.66406	0.28416	120916.00	5.17667
1958:1	4754.66406	0.28618	120904.00	4.12333
1958:2	4952.66406	0.28885	123540.00	2.91000
1958:3	5167.00000	0.28769	125176.00	2.60333
1958:4	5301.33203	0.28751	126812.00	2.99667
1959:1	5249.66406	0.29242	126968.00	3.90000
1959:2	5178.00000	0.29354	128336.00	5.03667
1959:3	5163.33203	0.29384	129732.00	6.10333
1959:4	5101.00000	0.29466	130956.00	5.69000
1960:1	5133.66406	0.29654	133568.00	4.91667
1960:2	5225.00000	0.29756	131000.00	3.98000
1960:3	5251.66406	0.29741	133000.00	3.10667
1960:4	5342.66406	0.29712	133440.00	3.98333
1961:1	5441.33203	0.29816	132384.00	3.69667
1961:2	5413.00000	0.29760	136076.00	3.53000
1961:3	5551.00000	0.29888	138464.00	3.13667
1961:4	5638.66406	0.29980	140732.00	3.10667
1962:1	5631.00000	0.30042	144732.00	3.46000
1962:2	5652.00000	0.30229	144892.00	4.10000
1962:3	5642.66406	0.30434	147100.00	5.57667
1962:4	5846.33203	0.30444	149732.00	4.40000
1963:1	5888.33203	0.30611	150704.00	4.15333
1963:2	6051.33203	0.30761	152568.00	3.64333
1963:3	6069.33203	0.31026	154092.00	4.06333
1963:4	6105.33203	0.31241	159532.00	4.18333
1964:1	6237.66406	0.31240	162548.00	4.11000
1964:2	6326.66406	0.31644	163216.00	4.01333
1964:3	6350.66406	0.31910	165676.00	4.23667

Period	$M1$	P	GDP	R
1964:4	6426.00000	0.32098	166576.00	4.43667
1965:1	6532.33203	0.32244	171692.00	4.42667
1965:2	6696.00000	0.32693	173368.00	4.71000
1965:3	6841.66406	0.32991	175916.00	5.15333
1965:4	6878.00000	0.33257	180460.00	5.78333
1966:1	7036.66406	0.33772	184592.00	6.04333
1966:2	7150.66406	0.34236	187100.00	6.14667
1966.3	7219.66406	0.34646	187624.00	6.36667
1966:4	7428.33203	0.34865	189736.00	6.51667
1967:1	7699.66406	0.35392	189332.00	5.90000
1967:2	7887.00000	0.35739	193132.00	5.23333
1967:3	7926.66406	0.35853	193972.00	5.73000
1967:4	8074.66406	0.36326	194572.00	6.52333
1968:1	8051.00000	0.36814	196556.00	6.97333
1968:2	8031.00000	0.36980	201568.00	7.28000
1968:3	8376.00000	0.37219	205088.00	6.53333
1968:4	8545.00000	0.37517	209076.00	6.50333
1969:1	8726.00000	0.38165	210604.00	6.92667
1969:2	8910.66406	0.38729	211748.00	7.34333
1969:3	8821.00000	0.38967	214584.00	8.32333
1969:4	8871.33203	0.39344	218848.00	8.80000
1970:1	8916.33203	0.40107	218504.00	8.44000
1970:2	8955.33203	0.40471	217400.00	7.63000
1970:3	9088.00000	0.40701	221176.00	7.34000
1970:4	9252.00000	0.41113	220912.00	5.95667
1971:1	9609.66406	0.41356	222884.00	4.74000
1971:2	10039.00000	0.41741	229796.00	3.96667
1971:3	10406.00000	0.41955	236684.00	4.70333
1971:4	10818.66406	0.42546	239184.00	4.63333
1972:1	11096.66406	0.43421	238420.00	4.79000
1972:2	11396.00000	0.43861	244928.00	5.71333
1972:3	11826.00000	0.44457	245864.00	4.84667
1972:4	12401.00000	0.45242	252552.00	5.05667
1973:1	12795.00000	0.45947	260360.00	5.08000
1973:2	13238.00000	0.47399	262716.00	6.62667
1973:3	13622.33203	0.48851	263604.00	8.45667
1973:4	13901.66406	0.50431	270796.00	9.65000
1974:1	14225.66406	0.52434	273488.00	9.02667

Period	$M1$	P	GDP	R
1974:2	14918.33203	0.54486	275072.00	11.43667
1974:3	14655.00000	0.56133	276736.00	11.48333
1974:4	14739.00000	0.57342	278728.00	10.07667
1975:1	15629.66406	0.58375	278432.00	6.86333
1975:2	16238.33203	0.59664	281048.00	7.33667
1975:3	16850.33203	0.61345	285228.00	8.37667
1975:4	17855.66406	0.62827	288040.00	9.16333
1976:1	17700.66406	0.63983	294624.00	9.29000
1976:2	17803.00000	0.65702	301344.00	9.37667
1976:3	18102.00000	0.66058	303416.00	9.38333
1976:4	18331.00000	0.67543	303168.00	8.63667
1977:1	18767.00000	0.68344	308204.00	7.90000
1977:2	19350.00000	0.69604	309360.00	7.25000
1977:3	19696.00000	0.70570	311344.00	7.33667
1977:4	20208.66406	0.71216	317108.00	7.41667
1978:1	20686.33203	0.72114	320172.00	7.45333
1978:2	21110.33203	0.73222	324836.00	8.33667
1978:3	21691.33203	0.74804	327040.00	9.04667
1978:4	22400.33203	0.76456	330956.00	10.49667
1979:1	22175.00000	0.77947	334800.00	11.13333
1979:2	22841.00000	0.80861	336708.00	11.16667
1979:3	23461.00000	0.82639	340096.00	11.80000
1979:4	23427.33203	0.84863	341844.00	14.18333
1980:1	23811.00000	0.86693	342776.00	14.38333
1980:2	23612.33203	0.88950	342264.00	12.98333
1980:3	24543.00000	0.91553	340716.00	10.71667
1980:4	25638.66406	0.93743	347780.00	14.53333
1981:1	25316.00000	0.96523	354836.00	17.13333
1981:2	25501.33203	0.98774	359352.00	18.56667
1981:3	25382.33203	1.01314	356152.00	21.01666
1981:4	24753.00000	1.03410	353636.00	16.61665
1982:1	25094.33203	1.05743	349568.00	15.35000
1982:2	25253.66406	1.07748	345284.00	16.04999
1982:3	24936.66406	1.09666	343028.00	14.31667
1982:4	25553.00000	1.11641	340292.00	10.88333
1983:1	26755.33203	1.12303	346072.00	9.61667
1983:2	27412.00000	1.13395	353860.00	9.31667
1983:3	28403.33203	1.14721	359544.00	9.33333

Period	$M1$	P	GDP	R
1983:4	28402.33203	1.16059	362304.00	9.55000
1984:1	28715.66406	1.17117	368280.00	10.08333
1984:2	28996.33203	1.17406	376768.00	11.45000
1984:3	28479.33203	1.17795	381016.00	12.45000
1984:4	28669.00000	1.18438	385396.00	10.76667
1985:1	29018.66406	1.18990	390240.00	10.51667
1985:2	29398.66406	1.20625	391580.00	9.66667
1985:3	30203.66406	1.21492	396384.00	9.03333
1985:4	31059.33203	1.21805	405308.00	9.01667
1986:1	30745.33203	1.22408	405680.00	11.03333
1986:2	30477.66406	1.22856	408116.00	8.73333
1986:3	31563.66406	1.23916	409160.00	8.46667
1986:4	32800.66406	1.25368	409616.00	8.40000
1987:1	33958.33203	1.27117	416484.00	7.25000
1987:2	35795.66406	1.28429	422916.00	8.30000
1987:3	35878.66406	1.29599	429980.00	9.30000
1987:4	36336.00000	1.31001	436264.00	8.70000
1988:1	36480.33203	1.32325	440592.00	8.61667
1988:2	37108.66406	1.33219	446680.00	9.13333
1988:3	38423.00000	1.35065	450328.00	10.05000
1988:4	38480.66406	1.36648	453516.00	10.83333

Notes:

The variables in columns 2 to 5 are are money stock (M1) in C$millions, implicit price deflator for GDP, 1981=100, gross domestic product at constant 1981 prices in C$millions and 90-day prime corporate rate.

BIBLIOGRAPHY

Akaike, H. (1969) 'Fitting Autoregressive Models for Prediction', *Annals of International Statistics and Mathematics,* **21**, 243-247.

Banerjee, A., Dolado, J.J., Hendry, D.F. and Smith, G.W. (1986) 'Exploring Equilibrium Relationships in Econometrics Through Static Models: Some Monte Carlo Evidence', *Oxford Bulletin of Economics and Statistics,* **48**, 253-277.

Banerjee, A., Lumsdaine, R.L. and Stock, J.H. (1992) ' Recursive and Sequential Tests of the Unit Root and Trend Break Hypothesis', *Journal of Business and Economic Statistics,* **10**, 271-287.

Barth, J.R. and Bennett, J.T. (1975) 'Cost-push versus Demand-pull Inflation: Some Empirical Evidence', *Journal of Money, Credit, and Banking,* **7**, 391-97.

Baumol, W. (1952) 'The Transactions Demand for Cash: An Inventory Theoretic Approach', *Quarterly Journal of Economics,* **66**, 545-556.

Beggs, J.J. (1988) 'Diagnostic Testing in Applied Econometrics', *Economic Record,* **64**, 81-101.

Ben–David, D. and Papell, D.H. (1992) 'The Great Wars, the Great Crash and the Unit Root Hypothesis', *mimeo,* University of Houston.

Bera, A. and McAleer, M.J. (1989) 'Nested and Non-Nested Procedures for Testing Linear and Llog-Linear Regression Models', *Sankhya,* **51**, 212-224.

Bernanke, B.S. (1986) 'Alternative Explanations of the Money–Income Correlation', in Brunner, K. and Meltzer, A. (eds) *Carnegie-Rochester Conference on Public Policy,* Autumn, 49-100.

Beveridge, S. and Nelson, C.R. (1981) 'A New Approach to Decomposition of Economic Time Series into Permanent and Transitory Components with Particular Attention to Measurement of the Business Cycle', *Journal of Monetary Economics,* **7**, 151-74.

Blanchard, O.J. and Quah, D. (1989) 'The Dynamic Effects of Aggregate Demand and Supply Disturbances', *American Economic Review,* **79**, 655-73.

Boothe, P. and Poloz, S.S. (1988) 'Unstable Money Demand and the Monetary Model of the Exchange Rate', *Canadian Journal of Economics,* **21**, 785-798.

Box, G.E.P. and Tiao, G.C. (1975) 'Intervention Analysis with Applications to Economic and Environmental Problems', *Journal of the American Statistical Association,* **70**, 70-79.

Breusch, T.S. (1978) 'Testing for Autocorrelation in Dynamic Linear Models', *Australian Economic Papers,* **17**, 334-355.

Breusch, T.S. and Godfrey, L.G. (1986) 'Data Transformation Tests', *Economic Journal,* (Supplement) **96**, 47-58.

Breusch, T.S. and Pagan, A.R. (1979) 'A simple Test for Heteroscedasticity and Random Coefficient Variation', *Econometrica,* **47**, 1287-1294.

Brodin, P.A. and Nymoen, R. (1992) 'Wealth Effects and Exogeneity : The Norwegian Consumption Function 1966(1)-1989(4)', *Oxford Bulletin of Economics and Statistics,* (Special issue), **54**, 431-457.

Brown, R.L., Durbin, J. and Evans, J.M. (1975) 'Techniques for Testing the Constancy of Regression Relationships Over Time', *Journal of the Royal Statistical Society,* Series B, **37**, 149-92.

Campbell, J.Y. and Perron, P. (1991) 'Pitfalls and Opportunities: What Macroeconomists Should Know About Unit Roots', NBER Macroeconomics Annual (Cambridge, Mass: MIT Press), 141-201.

Charemza, W.W. and Deadman, D.F. (1992) *New Directions in Econometric Practice* (London: Edward Elgar).

Chiang, A.C. (1984) *Fundamental Methods of Mathematical Economics* (New York: McGraw-Hill) 3rd ed.

Chow, G.E. (1960) 'Tests of Equality Between Sets of Coefficients in Two Linear Regressions', *Econometrica,* **28**, 591-605.

Christiano, L.J. (1992) 'Searching for Breaks in GNP', *Journal of Business and Economic Statistics,* **10**, 237-250.

Clements, M.P. (1989) 'The Estimation and Testing of Cointegrating Vectors'. *Applied Economics Discussion Paper,* No. 79, Institute of Economics and Statistics, University of Oxford.

Clinton, K. (1973) 'The Demand for Money in Canada, Some Single-Equation Estimates and Stability Tests', *Canadian Journal of Economics,* **6**, 53-61.

Cox, D.R. (1962) 'Further Results of Separate Families of Hypotheses', *Journal of the Royal Statistical Society,* Series B, **24**, 406-424.

Cuthbertson, K. (1985) *The Supply and Demand for Money* (Oxford: Basil Blackwell).

Davidson R. and MacKinnon, J.G. (1985) 'Testing Linear and Log-Linear Regressions Against Box-Cox Alternatives', *Canadian Journal of Economics,* **25**, 499-517.

Davidson, J.E.H., Hendry, D.F, Srba, F. and Yeo, S. (1978) 'Econometric Modelling of the Aggregate Time Series Relationship between Consumers' Expenditure and Income in the United Kingdom', *Economic Journal,* **88**, 661-692.

Davidson, R. and MacKinnon, J.G. (1981) 'Several Tests for Model Specification in the Presence of Alternative Hypotheses', *Econometrica,* **49**, 781-794.

Dhrymes, P.J. (1970) *Econometrics: Statistical Foundations* (New York: Harper and Row).

Dickey, D.A. and Fuller, W.A. (1979) 'Distributions of The Estimators For Autoregressive Time Series With a Unit Root', *Journal of the American Statistical Association,* **75**, 427-431.

Dickey, D.A. and Fuller, W.A. (1981) 'Likelihood Ratio Statistics for Autoregressive Time Series with a Unit Root', *Econometrica,* **49**, 1057-1072.

Dickey, D.A., Bell, W.R. and Miller, R.B. (1986) 'Unit Roots in Time Series Models: Tests and Implications', *The American Statistician,* **40**, 12-26.

Dickey, D.A. and Rossana, R.J. (1990) 'Cointegrated Time Series: A Guide to Estimation and Hypothesis Testing', *Manuscript,* Department of Statistics, North Carolina State University.

Dickey, D.A., Jansen, D.W. and Thornton, D.L. (1991) 'A Primer on Cointegration with an Application to Money and Income', *Federal Reserve Bank of St.Louis,* 58-78.

Dolado, J.J., Jenkinson, T. and Sosvilla-Rivero, S. (1990) 'Cointegration and Unit Roots', *Journal of Economic Surveys,* **4**, 249-273.

Drobny, A. and Hall, S.G. (1989) 'An Investigation of the Long-Run Properties of Aggregate Non-Durable Consumers' Expenditure in the U.K.', *Economic Journal,* **99**, 454-460.

Duck, N.W. (1992) 'U.K. Evidence on Breaking Trend Functions', *Oxford Economic Papers,* **44**, 426-439.

Durbin, J. (1954) 'Errors in Variables', *Review of the International Statistical Association,* **22**, 23-32.

Engle, R.F. (1984) 'Wald, Likelihood Ratio, and Lagrange Multiplier Tests in Econometrics', in Griliches, Z. and Intriligator, M.D. (eds) *Handbook of Econometrics* (Amsterdam: Elsevier), Vol. 2, 775-826.

Engle, R.F. and Granger, C.W.J. (1987) 'Co-integration and Error Correction: Representation, Estimation and Testing', *Econometrica,* **55**, 251-276.

Engle, R.F. and Yoo, B.S. (1987) 'Forecasting and Testing in Cointegrated Systems', *Journal of Econometrics,* **35**, 143-159.

Engle, R.F., Hendry, D.F. and Richard, J.F. (1983) 'Exogeneity', *Econometrica,* **51**, 277-304.

Feinstein, C.H. (1972) 'National Income, Expenditure and Output of the United Kingdom', in Stone, R. (ed.) *Studies in National Income and Expenditure in the United Kingdom* (Cambridge: Cambridge University Press), Vol. 6.

Fiebig, D. (1987) 'Diagnostic Checking in Practice: The Case of Outliers and Influential Data', *Econometrics Discussion Paper* , No. 87-01, University of Sydney.

Fountis, N.G. and Dickey, D.A. (1989) 'Testing for a Unit Root Nonstationarity in Multivariate Autoregressive Time Series', *Annals of Statistics,* **67**, 419-428.

Freedman, C. (1983) 'Financial Innovation in Canada: Causes and Consequences', *American Economic Review,* Papers and Proceedings, **73**, 101-106.

Friedman, M. (1956) 'The Quantity Theory of Money–A Restatement', in *Studies in the Quantity Theory of Money* (Chicago: University of Chicago Press).

Fuller, W.A.(1976) *Introduction to Statistical Time Series* (New York: Wiley).

Godfrey, L.G. (1988) *Misspecification Tests in Econometrics* (Cambridge: Cambridge University Press).

Godfrey, L.G. (1978) 'Testing for Higher Order Serial Correlation in Regression Equations When the Regressors Include Lagged Dependent Variables', *Econometrica,* **46**, 1303-1310.

Godfrey, L.G. (1978) 'Testing for Multiplicative Heteroscedasticity', *Journal of Econometrics,* **8**, 227-236.

Goldfeld, S.M. and Sichel, D.E. (1990) 'The demand for money', in Friedman, B.M. and Hahn, F.H. (eds) *Handbook of Monetary Economics* (North-Holland: Amsterdam) Vol. 1, 300-356.

Gonzalo, J. (1989) 'Comparison of Five Alternative Methods of Estimating Long-Run Equilibrium Relationships', *Discussion Paper,* No. 89-55, University of California at San Diego.

Gordon, R.J. (1982) 'Price Inertia and Policy Ineffectiveness in the United States, 1890-1980', *Journal of Political Economy,* **90**, 1087-1117.

Gordon, R.J. (1985) 'Understanding Inflation in the 1980s', *Brookings Papers on Economic Activity,* **85**, 263-99.

Gordon, R.J. (1988) 'The Role of Wages in the Inflation Process.' *American Economic Review,* **78**, 276-83.

Granger, C.W.J. (1986) 'Developments in the Study of Cointegrated Economic Variables', *Oxford Bulletin of Economics and Statistics,* **48**, 213-228.

Granger, C.W.J. and Newbold, P. (1974) 'Spurious Regressions in Econometrics', *Journal of Econometrics,* **14**, 111-120.

Granger, C.W.J. and Newbold, P. (1988) 'Some Recent Developments in a Concept of Causality', *Journal of Econometrics,* **39**, 199-211.

Gregory, A.W. and McAleer, M.J. (1983) 'Testing Non-Nested Specifications of Money Demand for Canada', *Canadian Journal of Economics,* **16**, 593-602.

Gregory, A.W. and MacKinnon, J.G. (1980) 'Where's My Cheque? A Note on Postal Strikes and Demand for Money in Canada', *Canadian Journal of Economics,* **13**, 683-687.

Hafer, R.W. and Jansen, D. (1991) 'The Demand for Money in the United States: Evidence from Cointegration Tests', *Journal of Money, Credit and Banking,* **33**, 155-68.

Hall, A. (1991) 'Testing for a Unit Root in Time Series with Pretest Data Based Model Selection', *mimeo,* North Carolina State University.

Hall, S.G. and Henry, S.G.B. (1989) *Macroeconomic Modelling* (Amsterdam: North-Holland).

Hallman, J.J., Porter, D.P. and Small, D.H. (1989) 'M2 per Unit of Potential GNP as an Anchor for the Price Level', *Board of Governors of the Federal Reserve System Staff Study,* No. 157.

Hallman, J.J., Porter, D.P. and Small, D.H. (1990) 'Is the Price Level Tied to the Stock of M2 in the Long Run?' *unpublished manuscript.*

Harvey, A.C. (1981a) *Time Series Analysis* (Oxford: Philip Allan).

Harvey, A.C. (1981b) *The Econometric Analysis of Time Series* (Oxford: Philip Allan).

Hausman, J.A. (1978) 'Specification Tests in Econometrics', *Econometrica,* **46**, 1251-1257.

Hendry, D.F. and Richard, J.F. (1982) 'On the Formulation of Empirical Models in Dynamic Econometrics', *Journal of Econometrics,* **20**, 3-33.

Hoffman, D. and Rasche, R.H. (1989) 'Long-Run Income and Interest Elasticities of Money Demand in the United States', *NBER Working Paper,* No. 2949.

Hoffman, D. and Rasche, R.H. (1990) 'Identification and Inference in Cointegrated Systems: A Synthesis of Recent Developments', *NBER Working Paper,* No. 3465.

Hylleberg, S. (ed) (1992) *Modelling Seasonality* (Oxford: Oxford University Press).

Jarque, C.M. and Bera, A.K. (1981) 'An Efficient Large Sample Test for Normality of Observations and Regression Residuals', *Working Papers in Economics and Econometrics,* No. 47, Australian National University.

Johansen, S. (1988) 'Statistical Analysis of Cointegration Vectors', *Journal of Economic Dynamics and Control,* **12**, 231-254.

Johansen, S. (1989) 'Likelihood Based Inference on Cointegration', *Lecture Notes for a Course on Cointegration,* Seminario Estivo di Econometria, Centro Studi Sorelle Clarke, Bagni di Lucca, Italy.

Johansen, S. and Lutkepohl, H. (1991) *Introduction to Multiple Time Series Analysis* (Berlin: Springer-Verlag).

Johansen, S. and Juselius, K. (1990) 'Maximum Likelihood Estimation and Inference on Cointegration with Applications to the Demand for Money', *Oxford Bulletin of Economics and Statistics,* **52**, 169-210.

Keil, M.W. and Richardson, W. (1990) 'A Comparison Among Partial Adjustment, Rational Expectations and Error Correction Estimates of the Canadian Demand for Money', *Journal of Applied Econometrics,* **5**, 273-291.

Koenker, R. (1981) 'A Note on Studentizing a Test for Heteroscedasticity', *Journal of Econometrics,* **17**, 107-112.

Kormendi, R.C. and Meguire, P. (1990) 'A Multicountry Characterization of the Nonstationarity of Aggregate Output', *Journal of Money, Credit and Banking,* **22**, 77-93.

Krasker, W.S., Kuh, E. and Welsch, R.E. (1983) 'Estimation of Dirty Data and Flawed Models', in Griliches, Z. and Intriligator, M.D. (eds) *Handbook of Econometrics* (Amsterdam: North-Holland) 656-696.

MacDonald, R. and Taylor, M.P. (1992) 'The Monetary Model of the Exchange Rate: Long-Run Relationships and Short-Run Dynamics', *Discussion Papers in Financial Markets,* No. 6, University of Dundee.

MacKinnon, J.G. (1991) 'Critical Values for Cointegration Tests' in Engle, R.F. and Granger, C.W.J. (eds) *Long-Run Economic Relationships: Readings in Cointegration* (Oxford: Oxford University Press).

MacKinnon, J.G. (1992) 'Model Specification Tests and Artificial Regressions', *Journal of Economic Literature,* **30**, 102-146.

MacKinnon, J.G., White, H. and Davidson, R. (1983) 'Tests for Model Specification in the Presence of Alternative Hypotheses: Some Further Results', *Journal of Econometrics,* **21**, 53-70.

Maddala, G. (1992) *Introduction to Econometrics* (New York: Macmillan) 2nd ed.

Maddison, A. (1982) *Phases of Capitalist Development* (Oxford: Oxford University Press).

McAleer, M.J. (1987) 'Specification Tests for Separate Models: A Survey', in King, M.L. and Giles, D.E.A. (eds) *Specification Analysis in the Linear model* (London: Routledge and Kegan Paul) 146-196.

McAleer, M.J., Pagan, A.R. and Volker, P.A. (1985) 'What Will Take the Con Out of Econometrics', *American Economic Review,* **75**, 31-43.

McCallum, B.T. (1987) 'The Case for Rules in the Conduct of Monetary Policy: A Concrete Example', *Bank of Richmond Economic Review,* **73**, 10-18.

Mehra, Y.P. (1977) 'Money, Wages, Prices, and Causality', *Journal of Political Economy,* **85**, 1227-44.

Mehra, Y.P. (1991) 'Wage Growth and the Inflation Process: An Empirical Note', *American Economic Review,* **81**, 931-937.

Microfit 3.0 (1991) *Users Manual* (Oxford: Oxford University Press).

Muscatelli, V.A. and Hurn, S. (1990) 'Cointegration and Dynamic Time Series Models', *Discussion Papers in Economics,* No. 9008, University of Glasgow

Nakamura, A. and Nakamura, M. (1981) 'On the Relationship Among Several Specification Error Tests Presented by Durbin, Wu and Hausman', *Econometrica,* **49**, 1583-1588.

Nelson, C.R. and Plosser, C.I. (1982) 'Trends and Random Walks in Macroeconomic Time Series: Some Evidence and Implications', *Journal of Monetary Economics,* **10**, 139-162.

Ng, S. and Perron, P. (1993) 'Unit Root Tests in Autoregressive–Moving Average Models with Data Dependent Methods for the Truncation Lag', *mimeo,* Université de Montréal.

Pagan, A.R. (1987) 'Three Econometric Methodologies: A Critical Appraisal', *Journal of Economic Surveys,* **1**, 3-24.

Pagan, A.R. and A.D. Hall (1983) 'Diagnostic Tests as Residual Analysis', *Econometric Review,* **2**, 159-218.

Perman, R.J. (1989) 'Co-integration: An Introduction to the Literature', *Strathclyde Papers in Economics,* No. 89/7, University of Strathclyde, Glasgow.

Perman, R.J. (1991) 'Cointegration : An Introduction to the Literature', *Journal of Economic Studies,* **18**, 3-30.

Perron, P. (1988) 'Trends and Random Walks in Macroeconomic Time Series: Further Evidence From a New Approach', *Journal of Economic Dynamics and Control,* **12**, 297-332.

Perron, P. (1989) 'The Great Crash, the Oil Price Shock and the Unit Root Hypothesis', *Econometrica,* **57**, 1361-1401.

Perron, P. (1990a) 'Testing for a Unit Root in a Time Series with a Changing Mean', *Journal of Business and Economic Statistics,* **8**, 153-162.

Perron, P. (1990b) 'Further Evidence on Breaking Trend Functions in Macroeconomic Variables', *Econometric Research Program Memorandum,* No. 350, Princeton University.

Perron, P. (1992) 'Trend, Unit Root and Structural Change: A Multi-Country Study with Historical Data', *Proceedings of the American Statistical Association, Business and Economic Statistics Section,* 144-149.

Perron, P. (1993) 'The Great Crash, the Oil Price Shock and the Unit Root Hypothesis: Erratum', *Econometrica,* **61**, 248-249.

Perron, P. and Vogelsang, T.J. (1992a) 'Nonstationarity and Level Shifts with an Application to Purchasing Power Parity', *Journal of Business and Economic Statistics,* **10**, 301-320.

Perron, P. and Vogelsang, T.J. (1992b) 'Testing for a Unit Root with a Changing Mean: Corrections and Extensions', *Journal of Business and Economic Statistics,* **10**, 467-470.

Perron, P. and Vogelsang, T.J. (1992c) 'A Note on the Asymptotic Distributions of Unit Root Tests in the Additive Outlier Model with Breaks', forthcoming, *Revista de Econometria.*

Pesaran, H.M. (1974) 'On the general problem of model selection', *Review of Economic Studies,* **41**, 153-171.

Pesaran, H.M. and Deaton, A.S. (1978) 'Testing non-nested nonlinear regression models', *Econometrica,* **46**, 677-694.

Phillips, P.C.B. (1986) 'Understanding Spurious Regressions in Econometrics', *Journal of Econometrics,* **33**, 311-340.

Phillips, P.C.B. (1987) 'Towards a Unified Asymptotic Theory for Autoregression', *Biometrika,* **74**, 535-547.

Phillips, P.C.B. and Loretan, M. (1991) 'Estimating Longrun Economic Equilibria', *Review of Economic Studies,* **58**, 407-436.

Phillips, P.C.B. and Ouliaris, S. (1990) 'Asymptotic Properties of Residual Based Tests for Cointegration', *Econometrica,* **58**, 165-193.

Phillips, P.C.B. and Perron, P. (1988) 'Testing for a Unit Root in Time Series Regression', *Biometrika,* **75**, 335-346.

Pindyke, R.S. and Rubinfeld, D.L. (1991) *Econometric Models and Economic Forecasts* (New York: McGraw-Hill) 2nd ed.

Plosser, C.I., Schwert, G.W. and H. White (1982) 'Differencing as a test of specification', *International Economic Review,* **23**, 535-552.

Poloz, S.S. (1980) 'Simultaneity and the demand for money in Canada', *Canadian Journal of Economics,* **13**, 407-420.

Raj, B. (1992) 'International Evidence on Persistence in Output in the Presence of an Episodic Change', *Journal of Applied Economics,* **7**, 282-293.

Ramsey, J.B. (1973) 'Tests for specification errors in classical linear least squares regression analysis', *Journal of the Royal Statistical Society,* Series B, **31**, 350-371.

Rao, B.B. (1993a) 'Unit Root Hypothesis, New Classical and Keynesian Models', *Economics Letters,* **41**, 47-52.

Rao, B.B. (1993b) 'The Nature of Transactions in the U.S. Aggregate Goods Market', *Economics Letters,* **41**, 385-390.

Romer, C. (1989) 'The Pre-War Business Cycle Reconsidered: New Estimates of Gross National Product, 1869–1928', *Journal of Political Economy,* **97**, 1-37.

Said, S.E. and Dickey, D.A. (1984) 'Testing for Unit Roots in Autoregressive-Moving Average Models of Unknown Order', *Biometrika,* **71**, 599-607.

Sargan, J.D. and Bhargava, A. (1983) 'Testing Residuals from Least Squares Regression for being Generated by the Gaussian Random Walk', *Econometrica,* **51**, 153-174.

Schwert, G.W. (1987) 'Effects of Model Specification on Tests for Unit Roots in Macroeconomic Data', *Journal of Monetary Economics,* **20**, 73-103.

Schwert, G.W. (1989): 'Tests for Unit Roots : a Monte Carlo Investigation', *Journal of Business and Economic Statistics,* **7**, 147-160.

Sims, C.A. (1988) 'Bayesian Scepticism on Unit Root Economics', *Discussion Paper,* Institute of Empirical Economics, Federal Reserve Bank of Minneapolis, No. 3.

Sims, C.A. and Uhlig, H.(1988) 'Understanding the Unit Rooters: A Helicopter Tour', *Discussion Paper,* Institute of Empirical Economics, Federal Reserve Bank of Minneapolis, No. 4.

Steel, R.G.D. and Torrie, J.H. (1980) *Principles and Procedures of Statistics: A Biometrical Approach* (New York: McGraw-Hill) 2nd ed.

Stock, J.H. (1987) 'Asymptotic Properties of Least Squares Estimators of Cointegrating Vectors', *Econometrica,* **55**, 1035-1056.

Stock, J.H. (1993) 'Unit Roots and Trend Breaks', in Engle, R.F. and McFadden, D.(eds) *Handbook of Econometrics* (Amsterdam: North-Holland), Vol. 4.

Stock, J.H. and Watson, M. (1988) 'Testing for Common Trends', *Journal of the American Statistical Association,* **83**, 1097-1107.

Stock, J.H. and Watson, M. (1989a) 'Interpreting the Evidence on Money-Income Causality', *Journal of Econometrics,* **40**, 161-182.

Stock, J.H. and Watson, M. (1989b) 'A Simple MLE of Cointegrating Vectors in Higher Order Integrated Systems', *unpublished manuscript.*

Stockton, D.J., and Glassman, J.E. (1987) 'An Evaluation of the

Forecast Performance of Alternative Models of Inflation', *The Review of Economics and Statistics,* **69**, 108-17.

Theil, H. and Boot, J.C.G. (1962) 'The Final form of Econometric Equation Systems', *Review of the International Statistical Institute,* **30**, 136-152.

Urbain, J.P. (1992) 'On Weak Exogeneity in Error Correction Models', *Oxford Bulletin of Economics and Statistics,* Special issue on 'Testing Integration and Cointegration', **54**, 187-207.

Vogelsang, T.J. and Perron, P. and (1993) 'Additional Tests for Unit Roots Allowing the Possibility of Break in the Trend Function', *manuscript,* Université de Montréal.

White, H. (1980) 'A Heteroscedastic Consistent Covariance Matrix Estimator and a Direct Test for Heteroscedasticity', *Econometrica,* **50**, 1-25.

Wickens, M.R. and Breusch, T.S., (1988) 'Dynamic Specification, the Long Run and the Estimation of Transformed Regression Models', *Economic Journal,* (Supplement) **98**, 189-205.

Wu, D. (1973) 'Alternative Tests of Independence Between Stochastic Regressors and Disturbances', *Econometrica,* **41**, 73-750.

Zivot, E. and Andrews, D.W.K. (1992) 'Further Evidence on Great Crash, the Oil Price Shock and the Unit Root Hypothesis', *Journal of Business and Economic Statistics,* **10**, 251-270.

Index